IMPLEMENTING PLANNED MARKETS IN HEALTH CARE

STATE OF HEALTH SERIES

Edited by Chris Ham, Director of Health Services Management Centre, University of Birmingham

Current and forthcoming titles

Financing Health Care in the 1990s
John Appleby

Patients, Policies and Politics
John Butler

Going Private
Michael Calnan, Sarah Cant and Jonathan Gabe

Implementing GP Fundholding: Wild Card or Winning Hand?
Howard Glennerster, Manos Matsaganis and Patricia Owens with Stephanie Hancock

Controlling Health Professionals
Stephen Harrison and Christopher Pollitt

Public Law and Health Service Accountability
Diane Longley

Hospitals in Transition
Tim Packwood, Justin Keen and Martin Buxton

The Incompetent Doctor: Behind Closed Doors
Marilynn M. Rosenthal

Planned Markets and Public Competition
Richard B. Saltman and Casten von Otter

Implementing Planned Markets in Health Care
Richard B. Saltman and Casten von Otter (Eds)

Accreditation: Protecting the Professional or the Consumer?
Ellie Scrivens

Whose Standards? Consumer and Professional Standards in Health Care
Charlotte Williamson

IMPLEMENTING PLANNED MARKETS IN HEALTH CARE

Balancing Social and
Economic Responsibility

EDITED BY
Richard B. Saltman and Casten von Otter

Open University Press
Buckingham · Philadelphia

Open University Press
Celtic Court
22 Ballmoor
Buckingham
MK18 1XW

and
1900 Frost Road, Suite 101
Bristol, PA 19007, USA

First Published 1995

A catalogue record of this book is available from the British Library

ISBN 0 335 19426 5 (hb) 0 335 19425 7 (pb)

Library of Congress Cataloging-in-Publication Data
Implementing planned markets in health care: balancing social and
    economic responsibility/editors, Richard Saltman, Casten von Otter.
        p.   cm. — (State of health series)
    Includes bibliographical references and index.
    ISBN 0–335–19426–5   ISBN 0–335–19425–7 (pbk.)
    1. Medical care—Marketing.   2. Health care reform.
3. Competition.   4. Medical policy.   5. Medicine, State.
I. Saltman, Richard.   II. Otter, Casten von, 1941–   .   III. Series.
RA410.56.I47 1995
362.1'068'8—dc20   94–44777
                                                                  CIP

Typeset by Type Study, Scarborough
Printed in Great Britain by St Edmundsbury Press,
Bury St Edmunds, Suffolk

CONTENTS

Series editor's introduction		vii
Acknowledgements		ix
List of contributors		x

Introduction 1
Richard B. Saltman and Casten von Otter

Part I: The politics of contracting
1 Contracting and the purchaser–provider split 25
  Ray Robinson and Julian Le Grand
2 Contracting and solidarity: Market-oriented changes in
  Dutch health insurance schemes 45
  Aad A. de Roo
3 Regulation of planned markets in health care 65
  Göran Arvidsson
4 Contracting and political boards in planned markets 86
  Mats Brommels

Part II: Balancing Incentives and Accountability
5 Costs, productivity and financial outcomes of managed
  care 113
  Nancy M. Kane
6 Vouchers in planned markets 134
  Richard B. Saltman and Casten von Otter
7 Clinical autonomy and planned markets: The British
  case 156
  Stephen Harrison

Part III: Constructing Entrepreneurial Providers
 8   Self-governing Trusts and GP fundholders: The British
     experience   177
     Clive H. Smee
 9   Implementing planned markets in health services: The
     Swedish case   209
     Anders Anell
10   Competitive hospital markets based on quality: The
     case of Vienna   227
     Christian M. Koeck and Britta Neugaard

Part IV: Conclusion
11   Balancing social and economic responsibility   239
     Richard B. Saltman and Casten von Otter

Index   252

SERIES EDITOR'S INTRODUCTION

Health services in many developed countries have come under critical scrutiny in recent years. In part this is because of increasing expenditure, much of it funded from public sources, and the pressure this has put on governments seeking to control public spending. Also important has been the perception that resources allocated to health services are not always deployed in an optimal fashion. Thus at a time when the scope for increasing expenditure is extremely limited, there is a need to search for ways of using existing budgets more efficiently. A further concern has been the desire to ensure access to health care of various groups on an equitable basis. In some countries, this has been linked to a wish to enhance patient choice and to make service providers more responsive to patients as 'consumers'.

Underlying these specific concerns are a number of more fundamental developments which have a significant bearing on the performance of health services. Three are worth highlighting. First, there are demographic changes, including the ageing population and the decline in the proportion of the population of working age. These changes will both increase the demand for health care and at the same time limit the ability of health services to respond to this demand.

Second, advances in medical science will also give rise to new demands within the health services. These advances cover a range of possibilities, including innovations in surgery, drug therapy, screening and diagnosis. The pace of innovation is likely to quicken as the end of the century approaches, with significant implications for the funding and provision of services.

Third, public expectations of health services are rising as those who use services demand higher standards of care. In part, this is stimulated by developments within the health service, including the availability of new technology. More fundamentally, it stems from the emergence of a more educated and informed population, in which people are accustomed to being treated as consumers rather than patients.

Against this background, policy-makers in a number of countries are reviewing the future of health services. Those countries which have traditionally relied on a market in health care are making greater use of regulation and planning. Equally, those countries which have traditionally relied on regulation and planning are moving towards a more competitive approach. In no country is there complete satisfaction with existing methods of financing and delivery, and everywhere there is a search for new policy instruments.

The aim of this series is to contribute to debate about the future of health services through an analysis of major issues in health policy. These issues have been chosen because they are both of current interest and of enduring importance. The series is intended to be accessible to students and informed lay readers as well as to specialists working in this field. The aim is to go beyond a textbook approach to health policy analysis and to encourage authors to move debate about their issue forward. In this sense, each book presents a summary of current research and thinking, and an exploration of future policy directions.

Professor Chris Ham
Director of Health Services Management Centre
University of Birmingham

ACKNOWLEDGEMENTS

First drafts of the essays in this volume were discussed at a workshop hosted by the Swedish Institute for Work Life Research in Stockholm in September 1993. Two days of spirited debate played an important role in the revision of the essays into final form. The participants included Sven-Eric Bergman, Per Carlsson, Helen Darling, Finn Didrichsen, José Manuel Freire, Gunnar Gelin, Brad Gray, Anna Hedborg, Christina Kärvinge, Simo Kokko, Torleif Olhede, Ingrid Pettersson, Igor Sheiman and Ilkka Vohlonen. The editors are grateful to the Swedish Institute of Work Life Research for a grant in support of the workshop and this volume, to Anna Seth who coordinated the workshop arrangements, and to Charlotte Brandigi who helped prepare the final manuscript.

LIST OF CONTRIBUTORS

Anders Anell is Director of the Swedish Institute for Health Economics in Lund.

Göran Arvidsson is Research Director at the Centre for Business and Policy Studies (SNS) in Stockholm.

Mats Brommels is Professor and Director of the Programme in Health Services Management in the Department of Public Health of the University of Helsinki School of Medicine.

Aad A. de Roo is Professor and Director of the Health Strategy and Management Programme at the Tilburg Institute of Advanced Studies of Tilburg University.

Stephen Harrison is Senior Lecturer in Policy Studies and Project Leader of the Yorkshire Collaborating Centre at the Nuffield Institute of Health of the University of Leeds.

Nancy M. Kane is Lecturer at the Harvard University School of Public Health in Boston, Massachusetts.

Christian M. Koeck is Director of the Department of Organizational Development of the Vienna City Hospital Association.

Julian Le Grand is Richard Titmuss Professor of Health Policy at the London School of Economics and Professorial Research Fellow at the King's Fund Institute, London.

Britta Neugaard is Research Assistant in the Department of Organizational Development of the Vienna City Hospital Association.

Ray Robinson is Professor of Health Policy and Director of the Institute for Health Policy Studies at the University of Southampton.

Richard B. Saltman is Professor in the Department of Health Policy and Management at Emory University School of Public Health in Atlanta, Georgia.

Clive H. Smee is Chief Economic Advisor in the Department of Health in London.

Casten von Otter is Professor and Head of the Public Sector Research Programme at the Swedish Institute for Work Life Research in Stockholm.

INTRODUCTION
Richard B. Saltman and
Casten von Otter

ON THE ROLE OF COMPETITIVE MECHANISMS

Health reform is now part of the daily political landscape in much of
Europe. In nearly all countries, major substantive reforms have
either been introduced, been proposed, or are under serious discussion. The objectives of reform invariably begin with micro-
economic concerns about obtaining greater efficiency and productivity from health providers, and maintaining as much equity as
possible in the distribution of health care services. In the pre-
ponderance of cases, an intense debate continues to turn on the
introduction of 'competition' and 'market discipline' as the most
appropriate mechanisms by which to achieve central reform objectives.

These overall commonalities, however, mask the multiplicity of
forms that the concept of 'competition' can take on. Competitive
incentives can be brought to bear on different actors in the system –
physicians and nurses, managers, patients, insurers (Saltman and
von Otter 1992). Viewed organizationally, competition can be
focused on different sectors of the health system – the finance side,
the production side and/or the allocation mechanism that connects
finance to production. Markets can be designed to turn on different
types of competitive incentives – price, quality, market share.
Different market mechanisms can be incorporated into a competitive framework; for instance, patient choice (in which money
follows the patient) or negotiated contracts (in which patients
follow the money). Once a specific market mechanism is selected,
multiple design questions remain as to how that mechanism should
be configured. With patient choice, patients may be allowed to

choose their general practitioner, their health centre, their hospital, their attending specialist and/or their insurance carrier. Negotiated contracts may be universal or selective, 'hard' or 'soft'; constructed on block payment, cost per case or on price and volume; quality of care may or may not be incorporated. The relative balance between risk and security for purchasers and for providers may change with each negotiation.

There is thus no unitary notion of how a planned market in health care should be constructed. Rather, in the real world of health reform from which evidence in this volume is drawn, there are multiple alternative competitive arrangements that can be introduced into existing tax-financed and social insurance-financed health care systems. To be certain, the configuration selected by policy-makers in different countries depends on more than just economic preferences. Health policy-making is by nature inextricably embedded in the social, historical, political and cultural contexts of the broader society within which health services are delivered.

This extended and extensive process of system redesign in the health sector is occurring at the same time that a major transformation is underway in the organization and structure of commercial corporations, particularly in the manufacturing sector. Over the past several years, large companies have embarked upon a process to 're-engineer' themselves into more efficient and effective systems of production (Hammer and Champy 1993). They have sought to flatten the managerial hierarchy and shorten decision-making cycles through a series of measures that include: (1) *streamlining* production through increased use of computers, robots and automated machinery; (2) *customizing* production into smaller batches, to meet the specific needs of different customers and/or market niches; (3) *out-sourcing* sub-components through networks of suppliers utilizing just-in-time delivery, partial sub-component assembly, and occasionally sub-component design; and (4) *decentralizing* operational decision-making to entrepreneurial business units and quasi-autonomous production teams. Commercial corporations have also reversed the traditional product design cycle, so that it begins with an assessment of customer needs, which then leads back to production requirements and production technology (Douma and Schreuder 1992).

This industrial transformation is intended to exploit the revolution in computerized information systems to produce higher quality products more efficiently, to enhance customer satisfaction and, thereby, to improve corporate market share and long-term profitability. These changes, of course, also have substantial negative aspects as well. Many white- and blue-collar workers find themselves forced to accept early retirement or lay-offs. Those workers who remain often find themselves working substantially harder for a company which has little long-term commitment to their well-being.

Not surprisingly, this thorough-going reform on the supply side of commercial industry, with the attendant academic and media attention, has had a spillover effect on the reform process underway in the health sector, which in most countries has also been focused predominantly on supply- or production-side activities. The broad structure of reforms across Northern Europe in tax-financed health systems such as the UK, Sweden and Finland suggest a number of parallels between health sector reforms and re-engineered corporations. These include an emphasis upon entrepreneurial business units (restructuring hospitals as self-governing trusts and public firms), performance-related pay (particularly for physicians) and team-based production (inside health centres and hospital departments). There is also increased emphasis in some countries upon client satisfaction as expressed through patient choice of general practitioners (UK, Sweden) and/or hospital (Sweden). Moreover, in an intriguing version of the commercial sector's re-engineering strategy, experiments are underway that decentralize all or part of hospital budgets into the hands of primary care doctors (UK) and/or local primary care linked political boards (Dalarna and Stockholm Counties in Sweden; Finland).

As in commercial corporations, the effort to restructure authority and to encourage entrepreneurial behaviour reflects both the possibilities created by the computerized information revolution as well as demand from patients and third-party payers for higher quality medical services produced less expensively and in a more customized fashion. Again as in corporate re-engineering, there are problematic aspects not only for those employees made redundant, but also for those who keep their jobs (see pp. 12–15).

This comparison between private manufacturing and publicly financed health care services should, however, only be pushed so far. While private manufacturing companies produce commodities, health care providers produce a social good. More importantly, health care services are not manufactured in controlled isolation by an individual company; rather, they are produced as part of a panoply of human services such as education, social care and home care which must be coordinated by a publicly responsible body and which, furthermore, interact intersectorally with such areas as housing, nutrition and employment. Consequently, while reforms to commercial corporations can be evaluated in terms of their effect on financial profitability for shareholders, reforms in the health sector must be judged in terms of achieving a variety of public policy objectives, including the distribution of services among all members of society, the degree to which these services are of equal quality, the intersectoral consequences of health care decisions on housing and employment, and, most importantly, the impact of health reform not just on medical services but on the overall health gain of the population.

ASSESSING THE REFORMS THUS FAR[1]

The long-term ability of competition-derived health care reforms to achieve what is a complicated mix of social and health as well as economic objectives has yet to be determined. While the implementation of production-side reforms that began in the UK in 1991 and in Sweden in 1990 have now generated initial financial results, these reforms have not been in place long enough to allow for meaningful evaluation of all consequences of specific competitive mechanisms for broader social and health-related objectives. Discussions about the appropriate role for competitive mechanisms in health care become still more complex when the range of countries is broadened to include two of the three industrialized countries which, unlike most, have developed proposals that would either introduce competitive incentives on the finance side of the health system, such as the seven-year effort in the Netherlands, or would consolidate quasi-anarchic private sector finance-side competition into a tightly corporatized approach known as managed competition, as put forward by the 1993 Clinton Administration proposal in the USA (the third outlier in this regard is New Zealand; cf. Borren and Maynard 1994).

The chapters in this volume point to a number of successes based on the introduction of market-oriented mechanisms. They also highlight a number of less clear-cut and/or less positive outcomes. Efforts to test the usefulness of market-style incentives on the finance side of health systems have been hindered by the extensive delays experienced in the implementation of regulated competition in the Netherlands. Policy analysts have already accumulated considerable information, however, on the behaviour of private insurance markets in a number of countries. Based on the best available evidence, it would appear that market-style mechanisms have considerably more potential to contribute to enhanced efficiency, effectiveness and cost containment when utilized as an allocative mechanism or introduced on the production side of the system than they have when introduced on the finance side of the system (Saltman 1994). Thus far, no country has succeeded in structuring a competition-based market on the finance side of their system for their entire population while still maintaining a commitment to universal access to equal services and to cost containment.

Conservative national governments in the UK (1988) and Sweden (1994) dropped consideration of competitive financing reforms once their consequences for universality and expenditure control became apparent. The only convincing evidence to suggest that a satisfactory finance-side market is possible exists in the realm of economic theory rather than real-world practice (Reinhardt 1992). Aad de Roo's chapter on the Netherlands provides a case study of the complexities involved in designing such a market. Multiple private insurers have powerful incentives to shed rather than retain high-risk patients. Given the strength of this incentive, governments confront a considerably more difficult task in designing regulation that can adequately control opportunistic behaviour (i.e. adverse selection and selective disenrolment) than they do in directly regulating the activities of hospitals and physicians. Dutch efforts to design a risk-rating formula that also can sustain equity of access have thus far been unsatisfactory (van de Ven 1993). A Swedish multi-party commission on health policy, HSU 2000, arrived at a similar conclusion (SOU 1993:38). In addition, in a finance-side market, administrative and other transaction costs tend to be higher at the same time that a government's bargaining leverage against provider-generated costs becomes fragmented and weak. As *The Economist* – not always a proponent of governmental intervention – has noted (29 May 1993), 'government can pool risks and use its muscle to keep down costs better than any private' insurer can.

Conversely, competitive mechanisms do appear to work reasonably well as allocative mechanisms and on the production side of health systems. Although competitive tendering for hospital support services in the early 1980s in the UK demonstrated only minimal savings (Key 1988), initial results from current reforms focused on clinical services appear to be more positive. Clive Smee's chapter concludes that, in Britain, fundholding general practitioners appear to have obtained lower prices and a better service for their hospital patients. In Chapter 9, Anders Anell presents initial results from various Swedish counties that include higher provider productivity and the elimination of most patient queues, which, combined with tighter economic controls and technological innovation, has made it possible to shrink overall in-patient capacity.

Some of the most creative competition-based initiatives among health care providers have yet to be adequately evaluated. The transformation of publicly operated hospitals into various types of public firms, for example, or changes in the payment incentives for hospital specialists, are by their nature difficult to assess in the short run. Evaluation of these and other structural changes affecting administrative and transaction costs, and of their impact on quality and equity aspects of a health system, will require a longer term perspective.

It is important to note the broader context within which evidence regarding the effectiveness of competitive mechanisms has been collected. With the exception of finance-side competition in the USA, experience to date has been within countries in Northern Europe which have a long-term government commitment to universal access and relatively strong regulatory pressures. In countries with publicly operated health systems (UK, Sweden, Finland), most provider institutions are directly accountable to public officials, and competitive activities have been introduced within organizational environments that incorporate a wide variety of national standards, and which engage in substantial monitoring and evaluation. Thus, the body of evidence presented in the chapters that follow comes predominantly from highly constrained health systems, evidence which also is supported by Nancy Kane's analysis of the unfettered provider market in the USA.

It is noteworthy that the current period of health reforms includes strengthened regulatory efforts by national governments. While these have been most visible in the pharmaceutical sector, involving the introduction of reference pricing and/or positive lists, similar regulatory measures have been introduced in other sectors of health systems as well. Thus, although the introduction of competitive incentives is the defining characteristic of planned markets in health care, these market-derived reforms have been accompanied by reinforced and expanded regulatory measures as well.

Health reform experience to date has also served to highlight a variety of obstacles to the implementation of effective and meaningful change. While some of these obstacles emerged specifically in response to the introduction of market-style reform instruments, many of the dilemmas detailed in the chapters below apply more broadly to all health sector reform activities. The most visible type of obstacles to date have been the following.

1 *Reform has been hard to implement.* While this may not surprise political scientists (Pressman and Wildavsky 1973), slow and uneven progress has tended to encourage opponents and reduce the scope of policy outcomes. Partly because market-oriented reforms have taken health systems into uncharted waters, partly due to the less well-prepared learning-by-doing approach, partly due to the complexity of the health and medical care delivery process, and partly due to opposition from key health sector groups, most governments (with the exception of certain Swedish counties) have had to redesign key elements, defer introduction of major changes, and otherwise back and fill to try to maintain reform momentum. While these rapid policy mutations may have prevented major mistakes, they leave little time to think through likely problems and to develop barriers to perverse consequences. As Stephen Harrison notes in Chapter 7, progress has also been slowed by medical professionals worried about the effect of reforms on clinical autonomy, as well as by unions reflecting concerns about job security in the current economic environment.

2 *Market-oriented reforms have required increased although fundamentally restructured national regulation.* Recognition of the need for renewed regulation was somewhat belated in the first wave of reforms, and efforts to find market-channelling rather than market-constricting techniques have been difficult. Although it initially appeared counterintuitive to some in what has been a devolutionary, market-influenced stage of health system development, a new public sector regulatory apparatus is needed to monitor and evaluate the availability and quality of health services provided by increasingly autonomous local authorities and/or provider institutions. More controversially, once national policy-makers begin actively to monitor and evaluate service delivery, they typically tend to set specific standards for service quality and delivery in an effort to maintain parity for all citizens. To meet this need, emphasis is placed increasingly on

developing regulatory arrangements such as licensing, accreditation, inspection, and the linkage of reimbursement to preset performance levels. These new functions can be undertaken at a national level, although in some contexts regional evaluation has been found to be feasible.

New arrangements for monitoring, evaluating and setting service standards are increasingly accompanying the decentralization of the operating authorities in the countries discussed in this volume. In Sweden, the National Board of Health and Welfare has established a regional monitoring and evaluation process to review quality and outcomes in all twenty-six counties (Spri 1992). In Finland, a new national oversight committee was proposed to guarantee that every municipality will provide basic services (Finnish Ministry of Health and Social Affairs 1993). In the Netherlands, a 1992 bill on health care information stated that the government expected its monitoring and regulatory roles to increase as responsibility for service delivery is decentralized. In the UK, the 1989 White Paper pointedly noted that the Secretary for Health would retain reserve powers to overrule decisions of self-governing trusts and/or fundholding GP practices, and, as Smee notes below, central control has in fact grown stronger in a number of key decision-making areas. In a decidedly different health system context, the 1993 Clinton reform proposal in the USA emphasized the necessary role of national standards and regulation in an effective health reform process (Starr and Zelman 1993).

3   *Conflicts between different policy objectives (efficiency vs equity or effectiveness) and between different reform instruments (negotiated contracts vs patient choice) have confused implementation.* One key conflict between efficiency and equity has emerged in the debate about the contents of a 'basic package' for health services. Attempts to define such a package are a response to the continued expansion of clinical capabilities on the one hand, in the face of constrained public sector revenues for health programmes on the other. The notion of a basic package is essentially a routinized form of health care rationing, reconfigured as an administrative device.

A number of governments, including the Netherlands national government and the Oregon state government in the USA, have undertaken to define such a basic package. Finland has proposed to monitor the provision of 'basic services' by municipalities, and New Zealand has also proposed developing a basic package. In

Sweden, a national commission has been established to provide guidance on the question of prioritization in health care services. The notion of a basic package driven by cost-related and/or service rationing concerns has the potential to contravene equity-based principles of medically necessary service provision, depending upon how the package is designed and/or implemented. The issues involved in such a clash have been addressed in the 1992 Report of the Netherlands Committee on Choices in Health Care (Ministry of Welfare, Health and Cultural Affairs 1992), and by the report of the Swedish Priorities Commission, which rejected simplistic notions of rationing through administrative or other supra-professional rules.

Turning to conflicts among different reform instruments, the pursuit of performance through negotiated contracts has in certain circumstances undermined the maintenance of existing equity-based service criteria. In Sweden, as Anell mentions in Chapter 9, a conflict between publicly let contracts and the principle of patient choice of provider could well emerge. Other commentators worry that patient choice itself has induced market-style behaviour that threatens the carefully planned equity of access which was the hallmark of the previous command-and-control administrative framework. A third threat to access has been created by the preference of private specialists to locate only in the wealthiest urban sections of the country. In the UK, concerns were expressed during the first year after the 1991 reforms that patients in certain fundholding GP practices (which patients can choose to join) were receiving preferential treatment from hospital specialists. In response to this, the Secretary of State felt compelled to issue guidelines precluding preferential treatment for patients of fundholding practices.

4 *The costs of transition have been substantially higher than expected.* Financially, the introduction of competition-based reforms as allocative mechanisms and on the production side have proved expensive. To the extent that these and additional expenditures reflect new transaction costs associated with market-based reforms (contract negotiation and litigation, advertising, higher personnel salaries), as well as costs of national monitoring and evaluation activities, these higher costs have been recurring.

Philosophically and culturally, some reform instruments have been perceived as cutting back on collective responsibility for health services. Concerns of this character have focused in particular on competitive incentives on the finance side of systems, as well as on proposals to sell or lease publicly operated institutions to private operators. Efforts to introduce competitive mechanisms among publicly owned and operated hospitals also triggered debates about the consequences of 'privatization' for equal access of all citizens to high-quality medical care.

CONTRACT-BASED MODELS OF INSTITUTIONAL MANAGEMENT

The process of health care reform takes on a different but related complexion when one assesses its impact at the level of organizational behaviour and institutional management. In publicly operated health systems across Northern Europe, the administrative structure found itself unable to respond to new economic pressures resulting from shifts in the social environment and also in individual needs and wants. The problem was not that centralized command was able to control health care in too much detail, but rather that the whole system had grown increasingly out of anyone's direct control. The major stakeholders had reached an equilibrium where no-one effectively seemed to be in full command of the budget. Viewed from this perspective, planned markets appear as an attempt to find more effective administrative instruments, rather than as a conscious or intended surrender of political control over the main objectives of health policy to market forces.

Following this logical line, the new planned market models can be understood as agency systems (Clarke and McGuinness 1987). Agency theory focuses on the design of transaction-efficient systems, which, as Robinson and Le Grand discuss in Chapter 1, means finding the best way for 'principals' to reach their goals most effectively by designing the most solid contracts, in order to avoid 'opportunism', 'shirking' and 'rationality with guile' by their partners to the contract. Transaction cost analysis seeks to define the costs and benefits of either breaking up large hierarchies into purchaser–provider relationships ('buying'), or integrating producers in a hierarchical relationship ('doing'), mainly by analysing the monitoring and contracting ability of participants under the specific circumstances of purchase and/or employment contracting (Williamson 1975). Similar to re-engineering in the commercial manufacturing sector, new health care models need to take into account not only the advantages of scale related to technology and production, but also the alternative costs of governance.

A contract, whether regulating a work commitment or a purchase, can never be complete in every detail and is never without costs. In addition, the risks of opportunistic behaviour (or rent-seeking, or slack) need to be taken into account. The coordination of policy direction, entitlements and resources can often be improved for greater overall system effectiveness by the selective use of both market-like and administrative allocative instruments. As the increasing size of international corporations illustrates, hierarchy and planning are not necessarily disappearing in favour of markets. Rather, combinations of new market structures and linkages can be adapted to reorganize an overextended hierarchy into a network of vertically and horizontally coordinated institutions, managed by an unambiguous structure of incentives.

Equally important design challenges confront health system reforms with regard to mechanisms of monitoring and accountability. Accountability in planned market contracts is effectively based on market forces as well as administrative methods (i.e. formal liability and quality audit). However, as can be seen in Anell's description of the county-based Swedish reforms, high ethical standards and trust might often be the preferred solution to a contracting problem, superior also in terms of transaction cost efficiency (von Otter 1991). It is thus important that governance procedures are supportive of high standards in this respect. Among other factors, the level of ethical standards reflects the quality of the public and professional discourse, and with them the public's access to adequate information.

The 'principal–agent problem' discussed here defines the conditions of ideal formal and informal contracting – with delivery according to understood volume, quality and price. In organizational terms, the agency issue raises questions about how to:

- shape the best task-focused units from hierarchically integrated local centres and hospitals;
- develop forms of internal and external competition and contracts suitable to achieve health policy objectives;
- design a structure of accountability reflecting both qualitative and efficiency outcomes.

The complexity of the issue makes it necessary to find answers by trial-and-error rather than by theoretical analysis alone. This in turn gives support to the idea of introducing organizationally based 'self-structuring' systems, rather than theoretically-based blueprint models.

One dilemma with contract models reflects their potential impact on the structure of organizational power and accountability in health care institutions. While a central purpose of using contracts rather than traditional administrative hierarchies is to put the governors more firmly in command (see Chapter 7), some contract models can serve in practice to reinforce provider domination. Conversely, administrators as a class are seen by some analysts (e.g. the public choice school) as self-seeking budget maximizers, who succeed because other institutional actors are more dispersed and cannot exercise effective countervailing power. In a contract model, provider-contractors could seek to increase their power *vis-à-vis* administrators and managers by organizing more intensively than would patients and taxpayers.

IMPLICATIONS FOR HEALTH PERSONNEL

The chapters in this volume suggest that the health reform process in several countries has unleashed powerful organizational forces, leading to strong pressures on the staff that run provider institutions. The catalytic process has been, perhaps contrary to some expectations, both market-driven and managerial in nature.

The effects so far point at increased segmentation of health care workers (especially between acute and chronic sectors), leaner work teams, intensification of the work process, readjustments in the structure of qualifications, a new professional role for doctors with less clinical autonomy, and less variation in low-status jobs. The effects for health care workers include a transformed employment contract, with new regulations for work itself, wages, schedules, ethics, qualifications, job security, etc. In essence, health care workers are being asked to absorb more risk and variation in the regular work process.

In most European health care systems, approximately three-quarters of total health care costs are personnel-related. These expenditures become obvious targets for any austerity programme, as well as for strategies to increase competitiveness. 'Lean production' in health care involves not only lower levels of staffing and more intensive work, but also restructured working time arrangements, a new division of labour, and related issues of economic compensation.

In publicly operated health systems, the level of trade union

membership has generally been high. The conditions of employment and work have been regulated through a nationally uniform
system, although typically with separate units for physicians and
other health care professionals. This nationally negotiated model
now appears to be dissolving in stages, with new local variation
appearing in payment systems, co-determination, and permanent
(*vs* temporary) employment agreements regulating specific issues of
local interest. In this process, the trade unions are being marginal-
ized.

The changes represent, on the part of the employers, an effort to
increase competition on the supply-side for health care workers. It
is probably no exaggeration to claim that flexibility in the use of
labour, and in payment systems and levels, is one of the most sought
after effects of the entire health reform process. In Sweden, when
the ambulance service in one county was privatized, salaries were
cut by 10–15 per cent. Compensation levels for overtime, on-call,
night shift, etc., were reduced. (When the company began to make
shift changes between drivers on the road in the midst of emergency
calls to save overtime compensation benefits, however, their contract with the county was terminated.)

In summary, the following observations can be made about the
likely consequences of recent reforms for health sector employees:

1 A flexibility in labour regulations and related practices is being
  used to adapt compensation-levels to new realities in the health
  sector labour market. The level of integration of production has
  been decreased, making it easier to set different types of workers
  – bound by different contract conditions – against each other (e.g.
  municipal workers' union *vs* transportation workers' union).
  New independent suppliers have thus far competed mainly by
  introducing more cost-effective staffing, rather than by technical
  or other innovations.
2 Workers are being increasingly hired on a temporary basis.
  Workers often have to accept temporary job offers, a replacement job, or even work by-the-hour. Restrictions on sub-
  contractors inside health care organizations have been lifted,
  making it easier to take on workers on a short-term basis.
3 Working time is another strategic factor in reducing costs. In
  Sweden, one-third of the members of the public blue-collar union
  (SKAF) have had to accept a reduced work-day. Until recently,
  there was a long-term objective, successfully applied in an increasing number of hospitals, to reduce weekend duty to two out

of five weeks. Now it is back to three out of five. In several instances, a split-shift has been reintroduced (e.g. forcing employees to work mornings and evenings with free time in between), a schedule that is highly unattractive to most people.
4 Flexible working hours, in the sense that schedules are individually planned with high staffing when there is a lot to do and with people given time off even at short notice when there is less, has been introduced in some institutions (von Otter and Viklund 1993). Arrangements are sometimes made with a 'time bank', in which time is deposited or withdrawn at different 'exchange rates' depending on the patient volume in the clinic. Although the process of designing flexible work arrangements began in Sweden when there was an expected shortage of labour, they have been useful with slight adjustments in the new economic conditions in the early 1990s.

Overall, the managerial catchword in recent health sector reforms has been flexibility, with reduced labour costs as the subtext. This orientation reflects the general economic policy trend in much of Europe and the industrialized world. As a means to incorporate the health sector in this broader policy shift, market-oriented reforms have been applied. A study, based on US experiences with contracting for different government functions, concludes that relying upon private contractors for public functions has reduced costs through three principal techniques (Kettl 1993).

First, as private contractors are free of public sector rules and civil service requirements, they have more flexibility than public agencies. They can use incentive pay systems and have greater freedom to hire and fire workers. They employ more part-time workers, have less absenteeism, and use employees for more than one task. Second, private contractors tend to pay lower wages than public sector agencies. Third, and most important, contractors tend to pay their workers substantially lower fringe benefits, especially retirement benefits. The difference in fringe benefits is 'the largest difference between the government and the private contractor' according to the National Commission for Employment Policy (Kettl 1993).

Broadly viewed, the issue of obtaining a more flexible health sector labour force could provide a valuable case study of transaction cost analysis in health sector reform. The effort to place salaried public employees on a contract basis tends to have positive short-term economic effects, for example by reducing costs for publicly operated health service providers and thus potentially freeing up scarce health care funding for alternative purposes. Yet, the use of employee contracts, particularly when the contract is let to a private for-profit sub-contractor, results in lower staffing levels, lower wages and poorer retirement benefits, especially for lower level (and less well-off) support staff. Evaluations of the early 1980s contracting-out process for support functions in British hospitals suggested that the added cost to society of unemployment insurance alone was sufficient to negate much of the short-term financial efficiency obtained (Key 1988). Since other public sector institutions will remain responsible for the financial consequences of redundancies and inadequate retirement provisions, a broad view of social and political transaction costs might lead health policy analysts to a different conclusion about the long-term effectiveness and appropriateness of market-style commercial contracting approaches.

ASSESSING THE EVIDENCE

The synthesis of public policy and market mechanisms remains in a formative phase. The necessary balance of social and economic goals has not yet been struck, nor has there been sufficient experience about how to combine these two seemingly antithetical elements to move the planned markets process to a more stable stage of development.

While the mix of specific mechanisms – and in particular the balance between competitive and regulatory measures – will vary and shift, the overall dimensions of reform will probably remain fairly well within present parameters. While some key questions in today's debate are likely to become less interesting, others will gain in importance. The reform debate itself will most probably take two seemingly opposite tacks. At the level of specific system design, questions about the usefulness of competitive incentives, their appropriate configuration, and their financial impact within the health system, will become more data-driven and specific. The economic efficiency related advantages of certain competitive incentives, judging from the evidence presented in the chapters below about Sweden and the UK, will become quite clear (see also Jonsson 1994; Robinson and Le Grand 1994). Conversely, the present role of political ideology, both on the Left in opposing competitive incentives almost entirely, and on the Right in trumpeting them as a magic solution for all welfare state programmes, will become more difficult to sustain. Rather, the complexity and difficulty of managing markets in any sector of health systems – and especially on the finance side of the system – will become increasingly apparent.

While more technical discussion concerning the design and financial impact of competitive incentives becomes more precise among system designers, questions among the broader population about the consequences of health reforms for public sector welfare state issues – including equity, quality, access to services and implications for health status – will grow. There is concern among some public health professionals that planned markets will inevitably lead to the same result as would the introduction of unregulated private markets in health care (Dahlgren 1994). Issues of importance to economists will take a back seat in a public debate focused on questions of broader public policy concern. The discussion concerning planned markets can thus be expected to shift from its current obsession with economic and financial efficiency to more traditional welfare state issues of equity and solidarity.

This emerging emphasis on health sector outcomes as they affect citizens will also reinforce present tendencies in planned markets towards strengthened regulatory mechanisms. Robinson and Le Grand (Chapter 1), Smee (Chapter 8) and Anell (Chapter 9) all illustrate the importance of regulatory as well as competitive elements in present planned market structures. Deteriorating conditions of employees in the health sector, especially but not exclusively para-professional and auxiliary personnel, can be expected to produce growing demands for corrective governmental intervention both to safeguard professional status as well as to preserve socially acceptable working conditions. Recognizing the permanency of the process that is underway, health care unions and professional organizations (like those in manufacturing industry before them) have begun to retreat from opposing nearly all changes in employment and working conditions to seeking to improve the situation for those employees who remain. Such improvements will involve concerted efforts to obtain governmentally imposed regulatory as well as negotiated contract types of solutions. Mandated quality standards as well as standard packages for health services are also likely to be introduced.

Despite these shifts in direction and emphasis, however, if the chapters in this volume lead to one overarching conclusion, it is the extent to which planned markets in health care have become all but irreversible. Much as it is difficult to imagine re-engineered private firms giving up flat hierarchies and entrepreneurial business units, it is hard to envisage hospitals in Northern Europe surrendering their new status as independently managed public firms, or patients in Sweden (and, for patients of fundholding general practitioners, in Britain) willingly giving up their greater degree of provider choice to return to the constraints of a catchment-area-based, administratively controlled, service delivery system. The current emphasis on micro-efficiencies will in practice become a permanent part of the health policy landscape. Thus, the new degrees of freedom made possible by the computer-driven information revolution, and reinforced by parallel economic and democratic pressures, are unlikely to allow a return to the centralized hierarchies of 1960s style planning models. Since there will be no return to a nostalgic era of good feelings and central political control, the core challenge for publicly operated health systems in Northern Europe will be to ensure that planned markets find more effective ways to satisfy both social welfare as well as financially and economically based criteria of success.

THE BOOK

The chapters that follow approach the current status of planned markets in health care systems from a variety of different perspectives. The authors consider the accomplishments and dilemmas of planned market models presently implemented or proposed in the UK, Sweden, Finland, the Netherlands, Austria (Vienna) and the USA. Although each essay is rooted in a specific country experience, the authors push beyond national boundaries to illustrate the broader conceptual notions that underlie these reform efforts. It is these broader conceptual notions that form the intellectual core of the book and that hold the most value for health policy-makers and analysts in other, inevitably somewhat different, national and health system contexts.

The essays are arranged according to the key theoretical notions they highlight. Part I, 'The politics of contracting', provides an empirical as well as conceptual review of the political issues associated with the design and introduction of planned markets in health systems. The contribution by Ray Robinson and Julian Le Grand, based upon three years of experience with new production-side arrangements in Britain, examines some of the fundamental assumptions upon which the Thatcher government's reforms were constructed. They argue that the movement from hierarchical systems of control to quasi-market arrangements was based upon a contract process that incurs excessive transaction costs. A detailed transaction cost analysis might well conclude that the former hierarchical approach or one based on contestability and yardsticks rather than outright competition would be preferable.

Mats Brommels focuses on recent health reform experience in Finland and Sweden. He contends that the core political culture of the Nordic countries, with its reliance upon locally elected officials, complicates efforts to introduce a purchaser–provider split inside the publicly operated health system. These elected public officials refuse to give up effective control over providers, since they consider themselves politically accountable to the electorate for service delivery. Brommels believes that a properly planned market for health care, structured upon correct incentives, would fully separate political from provider interests.

In a broadly normative contribution, Göran Arvidsson probes the appropriate role of public sector regulation within a planned market for health care. In his view, regulation is required to stabilize a health system incorporating competitive incentives, enabling it to achieve objectives set by a 'principal'. After exploring alternative models for this regulatory role, he concludes that it is essential to differentiate clearly between regulation as opposed to legislation or command and control administrative rules if planned markets are to develop appropriately.

The other chapter in this section, by Aad de Roo, focuses on the impact that the introduction of a more competitive purchasing system has had on solidarity in the Netherlands' social insurance based health care system. De Roo argues that increased utilization of efficiency-oriented contracting has begun to erode the equity characteristics of the Dutch system, and that hard-won improvements in solidarity are eroding. He raises important questions about the feasibility of introducing competitive incentives in the finance of health care while maintaining equity of access and coverage.

Part II, entitled 'Balancing incentives and accountability', focuses on the trade-offs inherent in seeking to increase the efficiency with which health care providers deliver care. Issues of undesirable behaviour, risk avoidance and professional autonomy are examined as part of the managerial aspect in designing planned markets. The first chapter in this section explores the emerging characteristics of a series of managerial mechanisms that are collectively known by the term 'managed care'. Drawing on more than a decade of experience with aspects of managed care in the USA, Nancy Kane analyses the effectiveness of managed care instruments both individually and as a group. She concludes that, while managed care may represent an improvement over the traditional fee-for-service in the USA, systems of managed care have not been able to lower hospital expenditures per capita and that they are 'highly susceptible' to selecting lower risk individuals as subscribers. These conclusions are particularly important in that the objectives of 'managed competition', as put forward in the 1993 Clinton Administration proposal as well as in the 1987 Dekker Report in the Netherlands, are premised upon the capacities of competing systems of managed care (e.g. managed competition).

Approaching the issues of risk and incentive from a social welfare perspective, Saltman and von Otter explore the appropriate role of vouchers in planned markets in health care. In comparison with social sectors like food or housing, where vouchers meet a fixed need for goods that are physical in nature, the use of vouchers in the provision of a human service like health care raises complex issues of risk allocation. The chapter concludes that vouchers should be carefully constructed if they are to be used for health services, and that they need to be tightly regulated to ensure that their use satisfies national health policy objectives.

The last chapter in this section considers the impact of market-oriented reforms on the physician's clinical autonomy. Stephen Harrison contends that the new structural arrangements in Britain, unlike previous NHS reforms, have had a noticeable impact on clinical freedom. He suggests that the loss of clinical autonomy due to the introduction of a planned market is interlinked with a drive towards explicit managerially led rationing of health care services.

The final and frequently neglected subject of discussion concerns the translation of ideas into practice. Part III, 'Constructing entrepreneurial providers', demonstrates the extent to which, despite far-reaching objectives, the actual process of health system reform has been incremental and pragmatic. The three chapters in this section also highlight similarities and differences in the type of planned markets presently under construction in the publicly operated health systems of the UK, Sweden and the city of Vienna, Austria.

Clive Smee examines the two key institutional innovations in the NHS reforms that began in 1991, self-governing hospital trusts and fundholding general practitioners. The leading parties to health reform in the UK agree that self-governing hospitals and fundholding practices are here to stay. However, Smee interprets the empirical evidence, relative to expectations, to show fundholding to be more effective and more conspicuously innovative than trusts.

Anders Anell's chapter on Sweden explores the various health reform models underway in the different counties. Seeking to draw together their underlying communalities, he argues that apparent differences are largely variations upon a theme (e.g. a purchaser–provider split). He concludes by remarking on the degree to which the Swedish version of a planned market relies upon patient choice of provider even as it adopts outwardly antithetical market-oriented instruments such as contracting.

In the final chapter in this section, Christian Koeck and Britta Neugaard describe an experimental effort to establish a planned market among the twenty-seven publicly operated hospitals of the city of Vienna. Adapting mechanisms utilized elsewhere, this programme seeks to establish a quality-driven process of health care reform. By expanding the principles of Total Quality Management (TQM) into a competitive framework for health reform, the Vienna experiment suggests the diverse objectives which seemingly similar market-oriented mechanisms can be structured to pursue.

NOTE

1 Parts of this section are based on Saltman (1994).

REFERENCES

Borren, P. and Maynard, A. (1994). 'The market reform of the New Zealand health care system: Searching for the holy grail in the Antipodes'. *Health Policy*, 27, 233–52.

Clarke, R. and McGuinness, T.M. (1987). *The Economics of the Firm*. Oxford: Basil Blackwell.

Dahlgren, G. (1994). *Framtidens Sjukvårdsmarknader: vinnare och förlorare*. Stockholm: Natur och Kultur.

Douma, S. and Schreuder, H. (1992). *Economic Approaches to Organizations*. Englewood Cliffs, NJ: Prentice-Hall.

Finnish Ministry of Health and Social Affairs (1993). 'Health reforms in Finland'. Helsinki, mimeo.

Hammer, M. and Champy, J. (1993). *Re-engineering the Corporation: A Manifesto for Business Revolution*. New York: Harper Business.

Jonsson, E. (1994). *Har den s.k. Stockholmsmodellen genererat mer vård för pengarna – en jämförnade utvärdering.* Stockholm: Institutet för kommunal ekonomi, Stockholm University.

Kettl, D.F. (1993). *Sharing Power: Public Governance and Private Markets.* Washington, DC: Brookings Institution.

Key, T. (1988). 'Contracting out auxiliary services'. In R. Maxwell (ed.), *Reshaping the National Health Service*, pp. 65–81. London: Policy Journals.

Ministry of Welfare, Health and Cultural Affairs (1992). *Report of the Government Committee on Choices in Health Care.* Rijswick: Netherlands.

Pressman, J. and Wildavsky, A. (1973). *Implementation.* Berkeley, CA: University of California Press.

Reinhardt, U.E. (1992). 'The United States: Breakthroughs and waste'. *Journal of Health Politics, Policy and Law*, 17, 637–66.

Robinson, R. and Le Grand, J. (eds) (1994). *Evaluating the NHS Reforms.* London: King's Fund Institute.

Saltman, R.B. (1994). 'A conceptual overview of recent health care reforms'. *European Journal of Public Health*, 4, 287–93.

Saltman, R.B. and von Otter, C. (1992). *Planned Markets and Public Competition: Strategic Reform in Northern European Health Systems.* Buckingham: Open University Press.

SOU (1993). *Three Models for Health-Care Reform in Sweden.* A Report from the Expert Group to the Committee on Funding and Organization of Health Services and Medical Care (HSU 2000). Stockholm: Ministry of Health and Social Affairs.

Spri (1992). *The Reform of Health Care in Sweden.* Stockholm: Spri.

Starr, P. and Zelman, W.A. (1993). 'Bridge to compromise: Competition under a budget'. *Health Affairs*, (suppl.), 12, 7–23.

van de Ven, W.P.M.M. (1993). 'Regulated competition in health care: Lessons for Europe from the Dutch Demonstration Project!' Paper given at the *Annual Meeting of the European Health Care Management Association*, Warsaw, Poland, 29 June–2 July.

von Otter, C. (1991). 'The application of market principles to health care'. In D. Hunter (ed.), *Paradoxes of Competition for Health.* Leeds: Nuffield Institute, University of Leeds.

von Otter, C. and Viklund, B. (1993). *La Flexibilité dans l'amenagement du temps de travail. La Suède: Le Cas des Soins Medicaux.* Paris: OECD.

Williamson, O.W. (1975). *Markets and Hierarchies: Analysis and Anti-Trust Implications.* New York: Free Press.

PART I

THE POLITICS OF CONTRACTING

1

CONTRACTING AND THE PURCHASER–PROVIDER SPLIT

Ray Robinson and
Julian Le Grand

INTRODUCTION

Many countries in Europe and elsewhere are experimenting with, or are considering experimenting with, market-type mechanisms for health service delivery (Hurst 1992; Saltman and von Otter 1992). They include the Netherlands, Sweden, New Zealand, the UK and almost all the former Soviet-Bloc countries of Eastern and Central Europe. Variously known as planned markets, quasi-markets, internal markets and public competition, the proposed systems take a variety of forms. However, most involve a split between the purchasers and the providers of health services, and the consequent need for some kind of contracting relationship between purchaser and provider units. It is with this relationship that this chapter is primarily concerned.

The discussion takes place in the context of just one of these market-oriented reforms, those affecting the UK National Health Service (NHS). Despite this limited focus, it is hoped that there are more general lessons to be learned, lessons that would be of interest to other countries engaged in similar exercises. This is partly because the British reforms are among the furthest along the road of implementation. But they are also of interest because in some ways the change they embody is extreme: the 'old' NHS was more like a command economy than almost any other health system in Europe, and the subsequent embrace of the market in the 'new' NHS more passionate.

For the benefit of those who may not be familiar with them, the chapter begins with a brief description of the NHS reforms. It continues with a brief review of the transition from a hierarchical bureaucracy to a contract-based system within the public sector generally and the NHS in particular, followed by a summary of some of the empirical evidence on the experience of contracting during the early period of the reforms. The next section considers some of the complications arising from the form of contracting that has been developed to date. In particular, it draws on the transactions cost literature in order to pose questions about the most appropriate form of contractual arrangements for health care systems. Arising from this discussion, the penultimate section offers some thoughts about directions that contracting of this kind should take if it is to economize on transactions costs, while simultaneously offering an incentive structure which is designed to improve efficiency in the provision of services. In this connection, the ideas of yardstick competition and contestability are discussed. Finally, the concluding section briefly draws together the line of argument developed in the chapter.

THE BRITISH NHS REFORMS

The main elements of the British health service reforms introduced on 1 April 1991 have been described in detail elsewhere (see Ham 1992; Le Grand and Bartlett 1993; Tilley 1993; Robinson and Le Grand 1994). On the demand side of the market, the key decision-makers are district health authorities (DHAs) and general practice (GP) fundholders. DHAs are government-appointed purchasing commissions that have responsibility for assessing the health care needs of their resident populations and for commissioning a mix of secondary and community health services which best meets these needs. Over the last two years, there have been a number of amalgamations among DHAs – as they have relinquished their previous provider functions – and several initiatives designed to integrate DHA functions with those of other agencies responsible for the commissioning of primary care (family health service authorities) and social care (local authority social service departments). Some analysts see these moves as a precursor to the eventual development of unitary, perhaps elected, authorities responsible for commissioning primary, secondary and possibly community care (Audit Commission 1993).

Operating alongside DHAs on the demand side of the market are

GP fundholders. These GPs receive budgets top-sliced from district allocations with which they can purchase a range of diagnostic, outpatient and elective inpatient procedures for patients registered with them. The rate of growth of fundholding has proved to be one of the more unexpected elements of the NHS reforms. From small beginnings, when it was widely regarded as a 'bolt-on' to the main reform agenda, fundholding has now grown to a point where some researchers are arguing that it offers the scope for an alternative, more patient-sensitive purchasing arrangement than is offered by the prevailing district model (Glennerster *et al.* 1994; Glennerster, 1994).

As fundholding has spread, however, some concerns have been expressed about the existence of dual agencies with the responsibility for purchasing secondary care: one basing its decisions on individual patients' needs (GP fundholders), with the other seeking to base its decisions on population health needs (DHAs). Critics of fundholding argue that comprehensive service provision and the achievement of equity objectives require population-based purchasing. Against this view, supporters of fundholding contest the distinction between 'population' needs and the needs of patients, and point to the efficiency gains that have already been achieved through the greater responsiveness of providers to GPs with actual purchasing power.

Recent attempts to resolve the dilemma of dual responsibility for purchasing have sought to use administrative guidelines in order to integrate GP fundholders' purchasing plans more closely with those of DHAs. In some regions, responsibility for certain of the management functions governing fundholders has been devolved to DHAs. Elsewhere, a number of DHAs have set out to counteract the attractions of fundholding by devolving notional budgets to non-fundholding GPs in an effort to offer them the opportunity to influence purchasing decisions more directly without actually becoming fundholders. However, somewhat against this trend towards greater coordination of GP (both fundholders and non-fundholders) and DHA decisions, some existing fundholders have started to group together in consortia or 'superfunds'. With the appointment of superfund managers, they aim to undertake centrally many of the management functions of individual fundholders – including negotiations with providers – and thereby reap economies of scale.

Taken together, it is clear that a far more complex configuration of purchasing organizations has developed than was envisaged when the NHS reforms were announced originally. Not surprisingly, this plurality of purchasing has introduced a number of additional complications into the purchasing and contracting process.

In comparison with the demand side, the supply side of the market is more straightforward. Its key feature has been the establishment of NHS Trusts. These are quasi-independent, non-governmental organizations providing secondary and community health services. They are directly accountable to the Secretary of State for Health, and NHS Executive Regional Offices now monitor Trusts to operationalize this accountability. Compared with units that were directly managed by DHAs, the Trusts have greater autonomy and freedom of action. This autonomy includes the ability to set the pay and conditions of service of their workforce, to decide upon the size and skill mix of their staff, and to exercise some limited new freedoms in relation to capital expenditure. By April 1994, 95 per cent of hospital and community services in the UK were provided by NHS Trusts (Bartlett and Le Grand 1994).

Such are the essential features of the NHS quasi-market. Within this market, it is service contracts that constitute the essential link between purchasers and providers. They make clear what services are to be provided and the terms on which they are to be supplied. In an uncertain world, they also have the important function of clarifying risk-sharing arrangements which may become relevant in the face of unplanned events on either the purchaser or provider side. In short, in theory at least, contracting has replaced management hierarchy in the NHS as the principal instrument of policy implementation. This is in fact part of a wider phenomenon in the British public sector, which we must now briefly consider.

FROM HIERARCHY TO CONTRACT

Towards the end of the 1980s, the British Government took a new direction in its efforts to reshape the public sector (Le Grand and Bartlett 1993). An essential element of the new approach was a move away from hierarchical, or vertically integrated, forms of organization towards quasi-market models based upon purchaser–provider separation and contractual relationships. While contracting was not entirely new to the NHS – indeed, there had been elements of competitive tendering for ancillary and for some clinical services throughout the 1980s (Ascher 1987; Ranade and Appleby 1989) – what was new was the scope of the new arrangements.

There are a number of possible explanations for this shift.

However, as Glennerster and Le Grand (1994) have argued, the principal motivation was probably the need to find a response to the tension between more demanding consumers of public services and limited resources. That aim led in turn to two other objectives (Harrison 1993), objectives which, to some extent, conflict with each other. On the one hand, there was the intention to delegate more responsibility down the line of management so that lower level managers would be given greater scope for using their own initiative; on the other hand, there was the intention of exerting more control in order to ensure efficient performance. To some extent, the apparent conflict between these two objectives was resolved by the purchaser–provider split, which offered greater autonomy but within which supply-side competition was designed to exert pressure on providers for efficiency similar to the process held to operate in the private sector.

As far as the delegation of responsibility downwards was concerned, the case for pursuing this strategy was expressed in the government's White Paper, *Competing for Quality* (Department of Health 1991: 2), in the following terms:

> Greater competition over the past decade has gone hand in hand with fundamental management reform of the public sector. This means moving away from the traditional pyramid structure of public sector management. The defects of the old approach have been widely recognised: excessively long lines of management with blurred responsibility and accountability; lack of incentives to initiative and innovation; a culture that was more often concerned with procedures than performance. As a result, public services will increasingly move to a culture where relationships are contractual rather than bureaucratic.

Harrison (1993) also points out that, as well as reducing the burden of control that hierarchies impose, the move to contracting has placed greater emphasis on local choice and performance assessment. As far as choice is concerned, the internal market is based upon the assumption that purchasers will have choice between competing providers. Whether choice actually exists in a number of quasi-monopoly markets, and whether the DHA purchasing function allows choice to be extended down to the level of the individual patient, is far less clear. Certainly, the popularity and growth of GP fundholding appears to have been stimulated by GPs' desire to maintain freedom to refer patients to hospitals of their choice rather than to be constrained by monolithic DHA decisions. On the other hand, some effort has been made by DHAs to operate a system of extra-contractual referrals – whereby patients of non-fundholders can be referred to providers with whom the host DHA does not have a contract – with a view towards offering choice to individual patients with non-standard needs. On the issue of performance assessment, the contract model places considerable emphasis upon performance indicators and monitoring. In principle, contracts offer an explicit format with which standards of performance can be specified and assessed. Once again, though, whether information systems are sufficiently well developed to allow for accurate monitoring is questionable (Keen 1993).

The second objective for moving from hierarchies to contracting identified by Harrison (1993) is the desire of government to exert more control over providers to ensure that they operate in the public interest. At first sight, this may seem surprising, as it might be thought that more control could be exerted within a hierarchial system. However, with many public sector hierarchies, it is argued, lines of accountability have become overextended with the result that self-serving provider or service-led cultures have developed. In short, provider interests dominate rather than those of users. By separating responsibility for purchasing from responsibility for providing, it is intended that this hegemony should be broken down. Certainly, recent ministerial statements have reasserted the role of purchasers as champions of the people who are expected to drive the system (NHSME 1993). Whether this role is achievable will depend in part on how well purchasers actually reflect the interests of users; a perennial issue in the British system, which has never offered much power or discretion to users themselves. It will also depend, to a large extent, on the way that contracting is developed. It is to this subject that we now turn.

CONTRACTS: THE EARLY EVIDENCE

During the first two years of the quasi-market, contracts between purchasers and providers varied a good deal according to local circumstances. However, they were all based upon three main categories: block, cost and volume, and cost per case.

Under block contract arrangements, access to a defined range of services and facilities is provided in return for an annual fee. This form of contract was particularly suited to the first year of the reforms because the NHS Executive sought to avoid major up-heavals by requiring health authorities to pursue a policy of 'steady state'.

This guidance allowed new systems to be put in place but was designed to minimize changes in patient flows and patterns of service delivery. As such, block contracts were able to reflect – albeit in contractual form – levels and patterns of activity that were already taking place. They also had the important practical advantage of being the least demanding in terms of information requirements. As a result, many purchasers simply took out block contracts with their local providers which reflected levels of activity and funding that were previously supplied by the provider within the pre-reform, unitary health authority. At the same time, however, even block contracts were more specific than the old style arrangements in terms of the requirements placed upon providers.

Variations in activity around indicative volumes were expected to be one of the main problems in operating block contracts. Providers might fail to use capacity to the full or treat more cases than had been agreed and funded. To cope with this problem, most block contracts specified ceilings and floors which permitted some variation around the expected level of activity. If actual activity fell outside this range, cost-and-volume arrangements came into operation.

A cost-and-volume contract specifies that a provider will supply a given number of treatments or cases at an agreed price. It allows the service specifications to be made more specific than is generally the case with a block contract. Greater emphasis is placed upon services defined in terms of 'outputs' (i.e. patients treated) rather than in terms of 'inputs' (i.e. facilities provided). If the number of cases exceeds the cost-and-volume agreement, extra cases have usually been funded on a cost-per-case basis.

Cost-per-case contracts are defined at the level of the individual patient. Because they involve a considerable level of transactions costs, health authorities have mainly used cost-per-case contracts to fund treatments that fall outside of block or cost-and-volume contracts. Referrals by GPs to providers with whom districts do not have prospective contracts (i.e. 'extra-contractual' referrals) have been the main form of district purchasing covered by cost-per-case contracts. Many services bought by GP fundholders have also been covered by cost-per-case contracts.

Not surprisingly, the majority of contracts taken out by DHAs in 1991–92 were block contracts. Table 1.1 shows that 83 per cent of

Table 1.1   Number and value of contracts for acute services, 1991–92 (percentages in parentheses)

Type of contract	Number of contracts		Value (£m)	
Block	1131	(40.7)	4346.5	(60.4)
Block with ceiling and floor	1179	(42.4)	2434.5	(33.8)
Cost and volume	169	(6.1)	314.1	(4.4)
Cost per case	108	(3.9)	17.6	(0.2)
RHA agency contracts on behalf of DHA	191	(6.9)	85.9	(1.2)
Total	2778	(100.0)	7198.6	(100.0)

Source: NAHAT (1992). Based upon returns from 101 DHAs.

contracts for acute services, by volume, were in this form – either simple block contracts or contracts specifying ceiling and floor levels of activity. In terms of value, the dominance of block contracts was even greater, accounting for 94 per cent of the total value of contracted services. Moreover, the steady-state requirement meant that most contracts replicated existing patient flows. Thus, as Table 1.2 indicates, block contracts placed with providers within the purchaser's own district accounted for the bulk of services, in value terms, in 1991–92. The size of these contracts – at an average value of £11.5 million – meant that they represented nearly 80 per cent of the total value of contracted services, although they accounted for only 18 per cent of the total number of contracts.

As the steady-state requirements were relaxed, however, there were signs that some DHAs began to use the contracting system in order to change the mix of services they commissioned and the terms on which they received them. A national questionnaire survey of DHAs' intentions for 1992–93 carried out by NAHAT (Appleby *et al.* 1992) showed that 61 per cent of purchasers intended to terminate some of their existing contracts and that 71 per cent intended to contract with new providers. Another survey by NAHAT (Appleby *et al.* 1992) also showed that among providers in the West Midlands Region (the largest region in the country covering a population of 5.2 million people), more detailed cost-and-volume contracts were due to increase from 1.7 to 10 per cent of

Table 1.2   Providers of contracts for acute services, 1991–92 (percentages in parentheses)

Type of provider	Number of contracts		Value (£m)	
NHS provider within district	488	(17.8)	5610.1	(78.6)
NHS provider outside district	2161	(78.6)	1503.4	(21.1)
Private sector	13	(0.5)	1.5	(0.0)
Voluntary sector	67	(2.4)	9.4	(0.1)
Other	19	(0.7)	12	(0.2)
Total	2748	(100.0)	7136.4	(100.0)

Source: NAHAT (1992). Based upon returns from 101 DHAs.

the total between 1991–92 and 1992–93, with block contracts falling from 97 to 89 per cent of the total over the same period. The proportion of block contracts still remained large, however.

The rapid implementation of contracting encountered a number of problems, most notably the lack of good information. Thus when the same survey questioned DHAs about the difficulties they had experienced in the contracting process, obtaining accurate data on comparative costs, on patient flows and on GP referrals were all cited as major problems by over 50 per cent of districts. Monitoring performance was also proving difficult, with over 90 per cent of DHAs citing the late arrival and poor quality of information supplied by providers as a major source of difficulty.

The general picture to emerge from these early findings is of an erstwhile hierarchy-based service grappling with the problems of rapid implementation of a new style contracting system over an extremely tight timetable. Despite adherence to the steady-state requirement in the first year of contracting, a number of difficulties were encountered. Over time, however, expertise is increasing and a perceptible movement away from the coarse mechanism of block contracting is taking place. In the future, as information systems become more sophisticated, less reliance will be placed on block contracts. For example, as better quality cost information becomes available, price tariffs based on more refined costings can be expected. In this connection, the national case-mix office of the

NHS Executive has already made a software package of health-related groups (HRGs, the UK version of DRGs) available to NHS providers as a basis for costings.

But these developments are not without their own problems. As contracting becomes more precise, it may well lead to steeply rising transactions costs. The early evidence of cost-per-case contracting in connection with extra contractual referrals and GP fundholder patients already confirms this tendency. This raises questions about the most appropriate contractual arrangements for the longer term.

CONTRACTS AND TRANSACTIONS COSTS

It has already become clear that the contracting process has incurred substantial costs. Setting up systems for recording, costing and billing has involved large investments in information systems. Many of these items of expenditure are non-recurring and will not be a source of higher costs in the future. But other costs will persist, particularly those associated with continuing transactions between purchasers and providers. These will constitute an additional category of expenditure that was not incurred under the pre-reform, unitary system.

In fact, the possibility that excessive transactions costs may be a source of inefficiency for market or quasi-market mechanisms has attracted the attention of a number of economists over the years. In particular, the work of Williamson (1975, 1986) has sought to identify those factors which, if present, mean that market contracts will be expensive to write, complicated to execute and difficult to enforce. If these conditions apply, firms may choose to bypass the market and rely upon internal, hierarchical forms of organization instead. Hence, transactions that would otherwise have taken place in the market are dealt with internally through administrative processes. Put another way, management hierarchies and markets can be viewed as alternative methods of economic organization for dealing with transactions. The choice between them should depend upon their relative efficiency.

In his work, Williamson identifies three features which, taken together, can be expected to favour internal organization over market transactions. These are bounded rationality, opportunism and asset specificity. *Bounded rationality* means that decision-makers, while seeking to act in a rational manner, can only be expected to do so to a limited extent. The bounded nature of behaviour arises because the capacity for individuals to formulate and solve complex problems is necessarily limited. These limitations become particularly important when faced with uncertainty about the future. If it becomes very costly or impossible to identify all future contingencies, and to specify adaptations to them, it may be more efficient to replace contract arrangements with internal, hierarchical organizations.

Opportunism refers to behaviour whereby individuals can be expected to pursue their interests through devious means. They may seek to derive advantage from the selective or distorted disclosure of information, or from making false promises. Information may be manipulated in a strategic fashion and intentions may be misrepresented. The existence of opportunism means that uncertainty is introduced into contractual arrangements as neither party can rely on the other one honouring non-legally binding promises. In such a world, internal organization may be a more effective means of controlling opportunism. It permits additional incentive techniques to be developed in order to curb opportunistic behaviour. In the limit, this may be achieved by fiat.

Asset specificity arises when transactions require investment in assets – both physical and human – that are specific to these transactions. As such, the parties to a contract have a continuing interest in each other because the nature of the commodity being traded depends upon an ongoing supply relationship. This arrangement is the converse of a spot market, where deals are struck by anonymous buyers and sellers. With asset specificity, market competition is liable to break down, as existing suppliers will enjoy advantages in relation to new entrants.

Hence the transactions cost approach suggests that when bounded rationality, opportunism and asset specificity are all present, internal organization may be a more efficient method of economic organization than market-type contracting between separate units. In the context of market-oriented health service reforms, this consideration raises the obvious question: Will transactions involving health services display these characteristics?

On the first characteristic – that of bounded rationality – there seems to be little doubt that this applies to health services. The nature of health and social care is highly complex, with major areas of uncertainty regarding, *inter alia*, the cost of individual services, their quality and, most important of all, measures of their outcome.

Whether opportunism will be a problem is less clear. Health service provision is traditionally viewed as embodying a set of values, based upon professional ethics and caring, which might be expected to exclude self-seeking and opportunistic behaviour. On the other hand, it would be naive to suggest that the strategic pursuit of self-interest has not always represented an element of health service provider behaviour, whether through corporate or professional vested interests. Whatever else it achieves, it seems extremely likely that the introduction of a more market-based approach will increase the incidence of this behaviour, and hence the potential for opportunism.

Asset specificity is another characteristic which seems to apply with particular force to health care services. Few of these services correspond to the simple type of consumer good which allows a person to enter a store, choose an item from the shelf, pay for it and disappear into the anonymity of private consumption. Much health care is a continuous, or at least a long-term, process involving treatment by a variety of agencies in many different contexts. This is especially true of long-term care and the treatment of chronic conditions. Even in the case of elective surgery, however, there is a complex chain stretching from pre-admission assessment through inpatient or day case treatment to post-discharge care. All of these considerations suggest that continuity in relations between purchasers and providers is likely to be important.

Taken together, therefore, there are strong reasons for believing that the conditions highlighted by the transactions cost approach are present in health services. One interpretation of how this might be expected to influence the contracting process between purchasers and providers has been put forward by Bartlett (1991). As he points out, block contracts have been the dominant form of contract in the NHS in the short run. These specify an annual fee in return for access to a defined range of services. They are broad-brush in nature and do not endeavour to specify prices for every eventuality. For this reason, they are necessarily incomplete and subject to opportunism. In particular, Bartlett believes that, despite the creation of mechanisms for measuring performance, opportunistic behaviour could lead to reductions in the quality of service provision, to an overemphasis on prestige treatments, and to an increase in organizational slack in the form of increased perks and side payments to staff. These can all be expected to raise the cost of services above the efficient level.

All of these considerations may be taken to suggest that efforts to create a quasi-market with a separation of purchaser and provider functions might be misplaced. Paradoxically, the transactions costs approach seems to suggest that the pre-reform hierarchical structure within a unitary health authority may have been the more efficient organizational structure after all. However, before reaching this judgement, some additional considerations need to be taken into account. Most notably, there is the role to be assigned to incentives.

INCENTIVES AND PERFORMANCE

Much of the case for the kind of internal market introduced in the NHS in the UK rests upon the belief that supply-side competition between rival providers will be a source of increased efficiency. It is recognized that there may well be extra transactions costs. However, the government believes that efficiency gains in service delivery due to competition will more than offset these. Whether they will or not is an empirical question that cannot be answered with certainty at the moment. But the case for introducing an incentive structure for increased efficiency is a powerful one. Indeed, Williamson himself points out that the vertical integration of firms must take place within the context of a competitive market for their final products and also for capital funds. These conditions provide external checks on the firm's efficiency. Now it is precisely the absence of such checks which led to the proposal for an internal market in the first place (Enthoven 1985). Without separate purchaser and provider functions, there would be no scope for the counterpart of competition in a final product market.

How can these apparently conflicting requirements be resolved? How can incentives be preserved but, in the light of transactions costs, the most efficient form of purchaser–provider organization be devised? Consideration of the theoretical literature and available empirical evidence (usefully reviewed in Propper 1993a) suggests that one approach would be to encourage purchasers and providers to enter into longer term, contractual relationships rather than to view their task as one of making spot market deals. This would avoid excessive transactions costs. Certainly, modern approaches to industrial organization and marketing emphasize the essentially collaborative nature of purchaser–provider relations (Davies 1991). Collaboration is a prerequisite for the effective sharing of information. In practice, this takes place far more widely in the private sector than discussions within the NHS generally allow.

With a move towards longer term contractual arrangements,

competition *for* markets – at the time of periodic contract negotiation – replaces competition *within* markets. In other words, there
is competition through franchising. At the same time, however,
mechanisms for ensuring efficient behaviour over the duration of a
contract period are still required. Responsibility for monitoring the
performance of the market and undertaking remedial action where
necessary falls most logically to some form of regulatory agency
(Propper 1993b). At the time of writing, it is unclear how this role
will be performed within the NHS. But whatever organizational
structure is chosen, two particular concepts are likely to be relevant
to the ways in which regulatory agencies carry out their work;
namely, yardstick competition and contestability.

REGULATION, YARDSTICKS AND CONTESTABILITY

Yardstick competition is a device used by regulators which enables
them to encourage efficiency in monopoly industries (Vickers and
Yarrow 1988; Kay and Vickers 1990). It is a way of bringing firms in
distinct markets indirectly into competition with each other. In its
simplest form, it operates through the regulation of pricing policy.
Thus, when regulators lack information on individual firms' cost
structures, the price that any firm may charge is set equal to the
industry's average costs. Faced with this constraint, profit-maximiz-
ing firms will have an incentive to increase productive efficiency
because they will thereby be able to increase their profits through
cost reductions. Moreover, dynamic efficiency will also be encouraged, because if a firm discovers a new technology, it can reap
the benefits of this superior technology until other firms catch up.
Conversely, firms that do not catch up incur losses.

The yardstick competition approach is not without problems.
There may be special factors beyond management's control which
lead to higher costs in some firms. Regulators need to identify these
instances and take them into account. Experience suggests that
there will be a tendency for all firms to argue that they face special
circumstances. There may also be a tendency for firms to collude
and thereby avoid the need to cut costs if they perceive yardstick
competition as a 'zero-sum game'. But, possibly most importantly,
there may be a danger that firms facing price constraints will try to
retain profitability by reducing the quality of their products. If this

danger arises, regulators need to extend the yardstick to qualitative features.

How might these ideas be applied to long-term contracting within health services? In their contracts, purchasers and providers could be required to set a price for a particular service based on long-run average cost for that service in the country as a whole. Providers would be free to retain any surpluses they made on the contract; hence they would have an incentive to reduce their costs below the national average. Special prices could be negotiated for providers with above-average costs for reasons beyond management control; for instance, if the higher costs arise because of the geographical location of the provider (such as an inner-city site), then setting the yardstick price on a local basis might be more appropriate than on a national one.

If this kind of scheme were applied in the NHS in the UK, it would entail some significant changes. First, it could be used only for cost-per-case and cost-and-volume contracts; purchasers and providers would thus have to shift their contracting procedure towards these and away from block contracts. Second, Trusts are currently only allowed to price according to *their own* (short-run) average costs, and cannot retain their surpluses. They would now have to price according to the yardstick, and at the same time be given the freedom to retain surpluses. But all these changes in fact seem desirable in any case. The extensive use of block contracts gives free rein to opportunism. Instructions to price according to one's own average cost may be unenforceable, since the regulatory authorities do not have sufficient information properly to assess the costs for each individual provider; providers, particularly those in monopoly positions, may well inflate their costs when reporting to the regulator (Ferguson and Palmer 1994). Even more seriously, the inability to retain surpluses gives Trusts little incentive to respond to market signals of any kind. In short, the application of yardstick competition to the NHS would not only be of direct benefit in reducing transactions costs and encouraging competition, but would bring useful gains in its wake by refining the operations of the quasi-market in key areas.

In a recent paper, Dawson (1994) has claimed that the own-cost pricing regulation currently in force in the NHS is likely to fail. The NHS market is characterized by contestability, small numbers and a high proportion of fixed costs; and the experience of such markets in the private sector, she argues, shows that prices tend to be negotiated between purchasers and providers, tend to be unique to each transaction and are usually secret. Regulators, she claims, will be unable to enforce behaviour contrary to the underlying incentives generated in these kinds of markets.

Does this argument also apply to yardstick competition of the kind suggested here? We believe not. This is basically because under yardstick competition, regulators have a much easier task than under an own-cost pricing regime. In order to perform their monitoring role, they do not need to check the costs of each and every provider; they simply need to know the price charged in each transaction to compare it with the yardstick price. This price will be incorporated into the contract document; hence keeping it secret would be difficult if not impossible. The focus for competition would be on quality, not on price; and the incentive for providers would be to cut costs, not to negotiate different prices.

'Yardsticks' defining clear standards of performance could also be used to assess the quality of service of different providers, and thereby to restrict opportunistic behaviour. In recent years, NHS performance indicators have been developed for a similar purpose. However, it is unlikely that performance indicators in their present form – that is, measures based primarily on inputs and activities – will provide the information that purchasers require. Similarly, aggregate measures such as the 'efficiency index', through which purchasers are currently required to meet annual performance targets, are also likely to be inadequate (Appleby and Little 1993). Over time, new indices of performance will need to be developed for assessing the standards of service offered by purchasers and providers (in terms of its cost, quality and outcomes) and for seeking improvements in these standards if they fall behind those achievable elsewhere (Appleby *et al.* 1993).

Most improvements in standards should be achievable through negotiation and mutually agreed action between purchasers and providers. However, there may be some cases where, for example, a purchaser is unable to obtain the standard of service it believes reasonable from its main provider. In such cases, alternative pro-vider arrangements must be a feasible course of last resort. Un-fortunately, the reliance on a long-term contractual relationship with an existing provider may mean that there are no comparable providers in the local market area. A local monopoly may have grown up. This, however, is where the concept of contestability becomes relevant.

Unlike a competitive market, a contestable market does not require the actual presence of a competitor; rather, it is the threat of new entrants to the market that acts as the stimulus for existing firms to act efficiently (Baumol 1982). If existing providers allow their prices to rise too high, or the quality of their service deteriorates too far, new entrants may be attracted to the market in the belief that they will be able to out-perform existing providers. Even here, though, there is a problem. The provision of health care requires substantial investment costs in buildings and equipment, and also considerable expenditure on initial staffing costs. These are likely to constitute a significant barrier to entry for new providers to a market. And so outside of those services where patients can be expected to travel for treatment, the checks on existing providers offered by contestability may be muted.

One way out of this difficulty is suggested by Culyer and Posnett (1990: 36). They argue that:

> Although the idea of contestability refers normally to the ease of entry of new firms . . . it is useful also to extend the notion to that of 'managements' which can see similar opportunities. The industrial ownership structure of the NHS lends itself conveniently to this sort of strategy, whether the hospitals whose management may be put out to tender are under the direct control of health authorities or Trusts. The implication is, of course, that some managements may disappear altogether, others may be recycled (probably in reconstituted form) elsewhere in the system, while, for all incumbent managements, there will be a continuing threat of their replacement.

Thus, Culyer and Posnett envisage the possibility of management teams, and indeed other groups of health workers, bidding for franchises which would periodically be re-contracted by regional health authorities or the Department of Health.

How far it would be practicable for policy to proceed in this direction is a matter for debate. Obviously, problems would arise in the transition period from one management to another. There is also the question of asset management: Who would be responsible for asset maintenance and investment, returns from which overlapped the length of the franchise? And where would the competing management teams be found? We cannot answer these questions here. But the line of thought does open up a range of possibilities for encouraging efficiency among providers in erstwhile monopoly or quasi-monopoly situations.

CONCLUSIONS

This paper has argued that the movement from hierarchical systems of control to quasi-market arrangements based upon contractual arrangements has been a general one within the public sector in the UK during the 1980s. The NHS reforms involving the introduction of an internal market based upon purchaser–provider separation is one aspect of this trend. The government's view is that the traditional hierarchical approach lacked the appropriate incentives for efficiency and led to over-concern with procedures rather than performance.

At the same time, however, while this trend has been in progress, critics have pointed to the excessive transactions costs that are sometimes incurred as part of the contract process. In doing so, the work of economists such as Oliver Williamson has been drawn on to raise the possibility that the former hierarchical approach may actually have been more efficient overall. As far as the NHS is concerned, resolution of this issue clearly depends upon the collection of more empirical evidence about the costs and benefits of alternative systems of organization.

In the meantime, this chapter has sought to offer the outlines of a strategy which seeks to economize on transactions costs but retain incentives for more efficient performance. The essential features of this strategy are longer term contractual relationships between purchasers and preferred providers, backed up by yardstick competition and management franchising. In many ways, this is a natural development of the model of managed competition that has been developed in the NHS over the last three years.

ACKNOWLEDGEMENTS

We are grateful for comments and helpful suggestions from the editors and from other participants at the Stockholm Conference, at which the paper that formed the basis of this chapter was presented.

REFERENCES

Appleby, J. and Little, V. (1993). 'Health and efficiency'. *The Health Service Journal*, 103, 20–22.
Appleby, J., Little, V., Ranade, W., Robinson, R. and Smith, P. (1992).

Implementing the Reforms: A Second National Survey of District General Managers. Birmingham: NAHAT.

Appleby, J., Sheldon, T. and Clarke, A. (1993). 'Run for your money'. *The Health Service Journal*, 103, 22–4.

Ascher, K. (1987). *The Politics of Privatisation: Contracting out Public Services*. London: Macmillan.

Audit Commission (1993). *Their Health, Your Business: The New Role of the District Health Authority*. London: HMSO.

Bartlett, W. (1991). 'Quasi-markets and contracts: A market and hierarchies perspective on NHS reforms'. *Public Money and Management*, 11, 53–61.

Bartlett, W. and Le Grand, J. (1994). 'The performance of NHS Trusts'. In R. Robinson and J. Le Grand (eds), *Evaluating the NHS Reforms*. London: King's Fund Institute.

Baumol, W. (1982). 'Contestable markets: An uprising in the theory of industry structure'. *American Economic Review*, 72, 1–15.

Culyer, A. and Posnett, J. (1990). 'Hospital behaviour and competition'. In A. Culyer, A. Maynard and J. Posnett (eds), *Competition in Health Care*. London: Macmillan.

Davies, G. (1991). 'Picking a philosophy'. *The Health Service Journal*, 101, 20–21.

Dawson, D. (1994). *Costs and Prices in the Internal Market: Markets vs the NHS Management Executive Guidelines*. Centre for Health Economics Discussion Paper No. 115. York: University of York.

Department of Health (1991). *Competing for Quality*. London: HMSO.

Enthoven, A. (1985). *Reflections on the Management of the National Health Service*. London: Nuffield Provincial Hospitals Trust.

Ferguson, B. and Palmer, S. (1994). *Markets and the NHSME Guidelines: Costs and Prices in the NHS Internal Market*. Centre for Health Economics Discussion Paper No. 120. York: University of York.

Glennerster, H. (1994). *Implementing GP Fundholding*. Buckingham: Open University Press.

Glennerster, H. and Le Grand, J. (1994). 'The development of quasi-markets in welfare provision'. Paper presented to the *Conference on Comparing Social Welfare Systems in Europe*, Oxford, May.

Glennerster, H., Matsaganis, M., Owens, P. and Hancock, S. (1994). 'GP fundholding: Wild card or winning hand?' In R. Robinson and J. Le Grand (eds), *Evaluating the NHS Reforms*. London: King's Fund Institute.

Ham, C. (1992). *Health Policy in Britain*, 3rd edn. London: Macmillan.

Harrison, A. (ed.) (1993). *From Hierarchy to Contract*. Newbury: Policy Journals.

Hurst, J. (1992). *The Reform of Health Care: A Comparative Analysis of Seven OECD Countries*. Paris: OECD.

Kay, J. and Vickers, J. (1990). 'Regulatory reform'. In G. Majone (ed.), *Deregulation or Re-regulation? Regulatory Reform in Europe and the United States*. London: Pinter.

Keen, J. (ed.) (1993). *Information Management in Health Services.* Buckingham: Open University Press.

Le Grand, J. and Bartlett, W. (eds) (1993). *Quasi-markets and Social Policy.* London: Macmillan.

National Association of Health Authorities and Trusts (NAHAT) (1992). *Purchaser Survey.* Birmingham: NAHAT.

NHS Management Executive (1993). *Purchasing for Health.* London: NHSME.

Propper, C. (1993a). 'Quasi-markets, contracts and quality in health and social care: The US experience'. In J. Le Grand and W. Bartlett (eds), *Quasi-markets and Social Policy.* London: Macmillan.

Propper, C. (1993b). 'Quasi-markets and regulation'. In J. Le Grand and W. Bartlett (eds), *Quasi-markets and Social Policy.* London: Macmillan.

Ranade, W. and Appleby, J. (1989). *To Market, To Market: A Study of Current Trading Activities in the NHS and the Implications of the Government's Provider Market Proposals.* Birmingham: NAHAT.

Robinson, R. and Le Grand, J. (eds) (1994). *Evaluating the NHS Reforms.* London: King's Fund Institute.

Saltman, R. and von Otter, C. (1992). *Planned Markets and Public Competition.* Buckingham: Open University Press.

Tilley, I. (ed.) (1993). *Managing the Internal Market.* London: Paul Chapman.

Vickers, J. and Yarrow, G. (1988). *Privatization: An Economic Analysis.* Cambridge, MA: MIT Press.

Williamson, O. (1975). *Markets and Hierarchies: Analysis and Anti-trust Implications.* New York: Free Press.

Williamson, O. (1986). *Economic Organisation: Firms, Markets and Policy Controls.* Brighton: Wheatsheaf.

2

CONTRACTING AND SOLIDARITY: MARKET-ORIENTED CHANGES IN DUTCH HEALTH INSURANCE SCHEMES

Aad A. de Roo

TAX-BASED AND INSURANCE-BASED HEALTH CARE SCHEMES

Solidarity as a political concept refers to social security schemes that limit or eliminate the consequences of social inequalities. In health care politics, it refers primarily to schemes designed to eliminate inequalities in access to and quality of care, originating from differences in social status, level of income or health status. Two central characteristics of such schemes are income-dependent cost-sharing and the absence of selection based on health risk. These characteristics are not a part of conventional health insurance, offered in the marketplace to people not willing or able to participate in social security schemes.

Conventional commercial health insurance is not a social arrangement; rather, it is designed as an indemnity insurance. Its mission is not the management of social inequalities, but risk management. Financially, its goal is not to break even, but to maximize return on equity for shareholders. Two main characteristics are fixed premiums, differentiated according to health risk, and selection based on health status. Social health care schemes deliberately include people who are on a low income and/or who are bad risks. Indemnity insurers deliberately exclude these people or make them pay substantially higher fees.

Social health care schemes can be public or private in nature. The public ones are established by public (governmental) decision-making bodies and apply to every citizen or to specified social groups. They are embedded in social law and regulation, and their costs are covered by taxation. Private schemes are based on decision-making by social insurance organizations. Their application is restricted to members of specific targets groups which are embedded in rules formulated by the organizations, and their costs are covered by the fees levied.

However, the differences between the two types of schemes cannot be adequately understood in terms of a purely public–private distinction (van Mierlo 1991). Because of the wide-ranging social interests involved in the functioning of social health insurers, detailed governmental rules and guidelines are typically imposed, which render these formally private insurers into quasi-public agencies. In the discussion below, the two types of social arrangements will be labelled tax-based and social insurance-based schemes. They will be discussed together with commercial indemnity insurance.

Modern tax-based and social insurance-based schemes in developed countries have a history that extends back to the end of the nineteenth century. After the Second World War, most schemes developed in the context of the larger social security policies of the welfare state. Differences in content and institutionalization reflect the idiosyncracies of the socio-political culture of the countries concerned.

These idiosyncrasies are also reflected in differences in the national mix of social and indemnity arrangements. Tax-based schemes in countries like the UK, Italy, Greece, Spain and Portugal offer health services to every citizen, delivered by public health care agencies. In these countries, indemnity insurance is also available, which offers an alternative to public services, by paying for services rendered by private health care providers. Thus, these countries have two separate health care sectors, a public one which is open to everybody, and a private one with limited access. Ireland has a different mix. Here the tax-based scheme is not open to all citizens, and specific social groups are dependent on indemnity insurance. However, unlike the UK and other countries, Ireland has no separate private health care sector. Indemnity insurance pays for the services offered to customers by the public health care infrastructure.

Countries with social insurance-based schemes, like the Netherlands, Germany, Belgium, France and Luxembourg, have a mix of social health care arrangements and indemnity insurance. Here the social arrangements are offered by sick funds. In Germany and the Netherlands, membership of these funds is mandatory for particular social groups below a certain level of income. France, Belgium and Luxembourg also have mandatory arrangements for certain social groups, but without level of income as a limiting factor. Persons without access to the sick funds are dependent on indemnity insurance.

SOCIAL INEQUALITIES AND THE MIX OF SOCIAL ARRANGEMENTS AND INDEMNITY ASSURANCE

The existence in most countries of a mix of a social health care scheme and indemnity insurance complicates the realization of solidarity in health care. Although social health care schemes are a prerequisite for equal access to and quality of care, they are not in themselves sufficient. Equality is also dependent on the extent to which indemnity insurance can offer better health care than the social arrangements.

Tax-based schemes guarantee equal access to and quality of care for all citizens. However, this does not necessarily mean that the health services offered will be of a high standard. If they are not, those individuals with the necessary purchasing power will turn to indemnity insurance to satisfy their needs (and some will try to buy services directly, without the protection offered by insurance against financial health risks). The more the services offered by social health care schemes fall short of the high standards required (e.g. long waiting lists, outdated equipment, understaffing, etc.), the greater the market opportunities for indemnity insurers.

By implication, generic solidarity will only exist in situations in which tax-based schemes are sufficiently well-funded that they meet the same standards as indemnity insurance. Where this is not the case, inequalities will exist between individuals with conventional insurance and those relying on the social arrangements. The gap between the two groups will depend on the level of access and standards of quality realized under the social health care scheme. Thus the level of health services established by the social arrangements will determine the probable range of inequality. Only when the social arrangements offer optimal care will the inequalities disappear.

A striking difference between Sweden and the UK may be explained from this perspective. The finance generated by the tax-based health care scheme in Sweden is large enough to provide a high standard of health care for everyone. Such an arrangement provides a very strong barrier to entry by indemnity insurance. In contrast, the UK tax-based scheme is unable to meet the necessary high standards of access and quality. The gap between what is possible and what is actually offered is large enough to create a substantial market in indemnity insurance for those people with the necessary purchasing power. This suggests that the volume of indemnity insurance in countries with a tax-based health care scheme can be used as a measure of the level of services offered by the social health care scheme.

Germany, the Netherlands, France, Belgium and Luxembourg have a mix of an insurance-based social health care scheme and indemnity insurance. The social arrangement is offered by sick funds and is mandatory for employees and persons dependent on social security payments. Such an arrangement provides for equality between participants because the income-dependent fees are not related to health risks, and the only criterion for entry is eligibility.

One key difference from tax-based schemes is that insurance-based schemes are not open to everyone. They are restricted to particular social groups identified in law or recognized by the sick funds themselves; everybody else is dependent on indemnity insurance. In most countries with a tax-based scheme indemnity insurance is a matter of choice for those with adequate purchasing power; in countries with an insurance-based scheme, there is no such choice. One of the consequences is that in these systems, particular low-income groups have no choice but to join an indemnity insurance scheme. This creates disproportional financial problems for these groups and permanently feeds discussions about how to improve the situation.

Of course, it is possible that members of these low-income groups refrain from purchasing health insurance at all, as is the case in the predominantly indemnity insurance system of the USA. It is interesting to observe, however, that in the Netherlands no such cases have been reported. This is consistent with the fact that health is very highly valued in the Netherlands, and thus people are prepared to allocate a considerable proportion of their income to health care. This may explain why low-income groups participate in indemnity insurance schemes that require disproportionate financial payments.

An additional explanation may be that indemnity insurance offers the possibility of substantially reducing premiums by limiting coverage to hospital expenditure. Premiums can be reduced further by accepting deductibles, so one has to cover a certain amount of the cost oneself. Such reduced-premium arrangements are very popular in the Dutch health insurance market. For low-income groups, however, they create a potential for health care underconsumption in terms of the financial barrier to access to primary care and dental care (not covered under such arrangements). And deductibles create incentives to postpone hospital care as long as possible.

Taken together in those countries with insurance-based schemes, the sick funds have the largest market share, reflecting the mandatory character of the arrangement (up to more than 60 per cent in the Netherlands and more than 70 per cent in Germany). As the schemes in Germany and the Netherlands do meet high standards of access and quality, in practice inequalities in health care are more or less eliminated. Again, this is not an inherent characteristic of the insurance-based approach. As in tax-based schemes, the decisive factor is the adequacy of funds for implementing the arrangement.

Over the long term, benefits in countries like the Netherlands, Belgium, France and Germany have paralleled advances in medical care. This has seen a rise in costs and a related rise in premiums. For macro-economic reasons, this development cannot go on indefinitely. Rising premiums can potentially lead to increased inflation, and so the governments in these countries have begun to develop policies to contain the growth in health care expenditure and in premiums.

The moment that premiums are unable to rise sufficiently to compensate for the additional costs incurred by medical advances, the existing equality of health care is put under pressure. The quality of care offered by the sick funds will begin to lag behind what potentially can be realized, creating or increasing the inequalities in care between the social health care scheme and that provided by indemnity insurance. This is exactly what is happening at the present time in the Netherlands. This chapter discusses the forces behind this development and looks at possible changes to the existing arrangements for countering inequalities in health care in the Netherlands.

The discussion begins with an overview of how solidarity became institutionalized in Dutch social insurance-based schemes. Second, the main Dutch market-oriented changes are reviewed: macro-budget, budgeting of insurers, replacing allocation by budgets with allocation by contracting, and the introduction of cost-related pricing. Third, the effects of these changes on equal access to and quality of acute hospital and ambulatory care will be evaluated. Finally, ongoing changes in the area of income-related cost-sharing will be considered.

SOLIDARITY IN AN INSURANCE-BASED HEALTH CARE SYSTEM: THE CASE OF THE NETHERLANDS

At the beginning of the twentieth century, the arrangements for health care in the Netherlands reflected the social structure of the country. There were three clearly discernible socio-economic groups. First, there were the poor, who had limited education, housing, social security and income. They had no regular work and so were dependent on charity. For them and for low-income workers, doctors provided consultations and home visits free of charge.

Second, many employees voluntarily joined one of the local sick funds which covered the services provided by doctors and pharmacists. In 1908, there were 616 such funds, 312 of which were cooperatives, 230 were managed by doctors and 74 were run for profit (van der Hoeven and van der Hoeven 1993). Doctors were highly supportive of these funds, as some payment was forthcoming for treating workers who traditionally had been treated for free. Also, the premiums the workers had to pay were small compared with those the well-off had to pay. For example, in 1900 doctors in Amsterdam offered their services to members of sick funds for an annual subscription of two guilders a year for a family of up to eight children; the well-off, however, were charged one guilder for each home visit (van der Hoeven and van der Hoeven 1993). Doctors earned most of their income from direct payments by the rich, those who did not have health insurance; they considered their work with the poor and members of sick funds a form of charity.

The situation at the beginning of the century fed several debates about the basic aspects of existing sick fund arrangements. The first concerned the for-profit character of some funds. Both politicians and the labour unions rejected profit-making as a legitimate aim of social policy. The doctors' association also opposed for-profit funds because profit-making was inconsistent with them reducing their fees to treat the poor. Eventually, for-profit sick funds disappeared from the scene, which culminated in licensing regulations banning them for good.

A second debate concerned the introduction of an income ceiling for admission to the funds. Doctors had a strong economic interest in excluding high-income earners, as they paid higher fees if they were outside the funds. This debate lasted until national legislation in the 1940s established an annually adjusted maximum income level.

A third debate concerned the position of those individuals with higher social status but a low income. For reasons of social status, many of them in the first decades of the century found it difficult to join a sick fund, yet at the same time they had difficulties paying the high premiums necessary for indemnity insurance. After the Second World War, social status was no longer taken into account, reflecting a development towards a more socially egalitarian society. However, the issue did not disappear completely from the agenda. Soon there was concern for the position of workers whose earnings rose above the income ceiling such that they had to terminate their membership of a sick fund. Subsequently, the discussion embraced what was to become of individuals on a high income – salaried or not – who experienced a substantial loss in income after retirement.

A fourth discussion concerned the voluntary character of health insurance. Although most workers participated in a sick fund, those who did not exposed themselves and their families to risks that were avoidable. In this debate, one side held the view that mandatory insurance was a way of patronizing the labour force, whereas the other side argued that such insurance was justified because some people (i.e. those unwilling to join a sick fund) had to be protected against themselves.

There was much controversy about how to solve these problems, and it was only under the pressure of the German occupation that most were settled by the Sick Fund Decree of 1941. This decree ruled that only non-profit-making sick funds could be licensed as health care insurers and it introduced mandatory participation for workers below a certain income level. The situation in which doctors got paid less for attending to sick fund patients remained. Health care rights were extended to hospital care and dental care and were offered for an income-dependent fee. Along these lines, a post-war sick fund scheme developed with important characteristics:

- income-dependent fees, thus creating solidarity between workers on different incomes, reflecting strong egalitarian social values;

- employers were to contribute to the payment of these fees on a 50:50 basis;
- the target group was limited to regular employed persons, to the exclusion of irregularly employed workers, workers with above-average salaries, workers on a pension, etc.;
- mandatory participation of members of the target groups (no health risk selection);
- implementation by local, semi-independent sick funds;
- nationwide cost-sharing between the sick funds and compensation for any operating deficits due to rising fees (open-end financing);
- no funding of deficits through taxes;
- coverage of the costs of primary care, hospital care, dental care and pharmaceuticals.

Though this social health care scheme was a major breakthrough, it only established solidarity among a substantial but restricted group of workers on a salary. It offered no solidarity between workers on a salary and low-income, self-employed individuals (such as shopkeepers, farmers, etc.). Neither did it provide solidarity between high-income and low-income employees, because, unlike in Germany, employees had to quit the sick fund if their income rose above the income ceiling (the difference between the indemnity insurance premium and the sick fund premium often far exceeding their rise in salary). And at the age of sixty-five, not only were employees forced to quit their sick fund, they also had to pay substantially higher indemnity insurance premiums at the very moment they were facing a substantial drop in income.

In reaction to these problems, two additional social insurance schemes were introduced after the Second World War that offered the same coverage as the mandatory sick fund – the voluntary sick fund for self-employed individuals and workers earning above the income threshold of the sick fund, and the voluntary sick fund for the elderly. Both funds used fixed premiums, thus doing away with an important solidarity feature of the mandatory sick fund – the relation between income and level of premium. But solidarity values were reflected in the fact that these funds rejected risk selection, accepting every applicant. Reflecting the same values, they did not offer deductibles on premiums if one accepted restricted reimbursement. Given their mix of indemnity and social insurance characteristics, they can be characterized as half-way arrangements. They also had a tax-based dimension, in so far as the government reluctantly began to compensate to a certain degree the operating deficits of these funds.

In this way, a post-war market segmentation developed based on socio-economic status. Social groups not eligible to join the mandatory sick fund or either of the 'half-way' health insurance schemes formed the third market segment, served by indemnity insurance. Indemnity insurers had been present in the Dutch health insurance market from the beginning. However, before the Second World War they did not play an important role, as their clients were in the main recruited from the relatively small high-status, high-income group that traditionally was accustomed to making direct payments to doctors. However, there was a modest growth in the indemnity insurance market in the 1950s and 1960s. As described above, these insurers select applicants on the basis of health status; if an individual is considered to be at high risk, they will be expected to pay higher premiums or be excluded from the scheme completely. They also offer deductibles and variation in coverage, ranging from full sick fund coverage to coverage limited to acute hospital care. These characteristics make such insurance schemes attractive for individuals with good health status (good risks) and an income that makes restricted reimbursement unproblematic.

POST-WAR PROBLEMS

The mix of a mandatory sick fund, two 'half-way' schemes and indemnity insurance was delicately balanced until the end of the 1970s. In the 1960s, the structure of Dutch society changed. One consequence was that socio-economic status gradually came to be viewed as less significant by those individuals who had a choice between indemnity insurance and the two 'half-way' schemes. As a rule before this change, workers without access to the mandatory sick fund joined the voluntary sick fund. Indemnity insurance was not a real alternative, as it was associated with a different socio-economic status.

Such a change in the importance of socio-economic status made possible consumer choice based on micro-economic calculations. This created new market opportunities for indemnity insurers to attract good risks from social groups excluded from the mandatory sick fund. Commercial insurers used these opportunities to substantially undercut the premiums of both 'half-way' funds, by offering policies with restricted reimbursement. Combined with rising health care costs during the 1960s and 1970s, this created growing operating difficulties for the voluntary fund and the fund for the elderly. The two funds were increasingly left with a membership whose incomes were below average but whose risk was above average. Inevitably, the health care costs of these funds grew disproportionately, which in turn led to a disproportionate increase in premiums. In relation to the already weak income position of the people concerned (especially the elderly and groups of independent workers), this was widely seen as an unacceptable loss of solidarity.

The first response to these developments was to minimize the disproportionate growth in premiums by expanding public funding of operating deficits. At the end of the 1970s, political opposition to this solution was substantial for two reasons. First, the government's budgetary problems left little opportunity for additional public funding; they actually created strong pressures to abolish any public funding of the two funds. Second, the growth in health care expenditure had become a major political issue. It was felt that such growth might increase inflation and therefore it had to be contained. Additional public funding of operating costs ran counter to this goal.

This led to a complicated situation at the beginning of the 1980s, in which macro-economic problems and social health insurance problems became interrelated. There was:

- over-representation of bad risks in the voluntary fund and the fund for the elderly, leading to increased operating deficits;
- diminishing political support for public funding of these deficits;
- an overall political desire to contain the growth of health care expenditure;
- political debate about compensation for the low-income members of both funds, who were affected by the disproportionate growth in premiums.

In 1986, the government closed down both the fund for the elderly and the voluntary sick fund. At the same time, a new 'half-way' social insurance scheme was introduced by the government, the Standard Insurance Scheme. It offered the same coverage as the mandatory sick fund for a nominal, income-independent premium fixed by the government. Entry selection and deductibles were not allowed, and indemnity insurers were obliged by law to offer the new insurance to former members of both the fund for the elderly and the voluntary fund. In effect, the government created a new 'half-way' arrangement and forced the indemnity insurers to run it.

The premium was based on average costs, which fell substantially short of the operating costs of the scheme (because of over-representation of bad risks). These extra costs were covered by a supplementary premium paid by all other subscribers of indemnity insurance. In this way, solidarity in terms of cost-sharing between good and bad risks was realized. But the level of solidarity remained limited, because the cost-sharing was based on premiums that were not income-dependent.

TOWARDS A MARKET-ORIENTED APPROACH

The introduction in 1986 of the Standard Insurance Scheme is generally referred to as the 'small system change'. It stopped the erosion of solidarity, caused by the attractiveness of indemnity insurance to 'good risks' not eligible to join the mandatory sick fund. And the closing down of the voluntary fund and the fund for the elderly meant the government was no longer obliged to use public money to cover their operating costs. But it offered no solution to the basic cost containment problem. Neither did it offer a solution to the disproportionate effects of an income-independent premium on the level of consumption of low-income groups. Politicians began to realize that once again they would have to face the problem of rising social health insurance premiums. Since the majority of Dutch hospitals are private voluntary institutions with their own strategic agendas, the seemingly most straightforward mechanism to achieve effective cost control was through a redesign of health financing arrangements. Therefore, the search for further reform continued, and this resulted in 1987 in the Dekker Report (see Commissie Structuur en Financiering Gezondheidszorg 1987). The Dekker Report proposed a radical break from the comprehensive planning framework that the government had introduced in the 1970s. This planning framework, which had cumulated in the 1982 Planning Act (Wet Voorzieningen Gezondheidszorg), was only partially implemented and was never fully accepted by Dutch hospitals and physicians. Its cost-containment effects were unclear (Maarse and Mur-Veerman 1990). The Dekker Commission advised the government to drop its comprehensive health care planning approach and to rely on market forces for effective cost containment of health care expenditure. The report contained several important proposals for restructuring existing health insurance arrangements. One proposal suggested replacing existing arrangements with two new schemes, one mandatory and one voluntary. The mandatory insurance scheme would cover long-term care and about 80 per cent of existing sick fund coverage for acute care, leaving other health risks to be covered by voluntary insurance or direct payments, whichever the individual preferred.

A related proposal was the introduction of a two-part premium, one part income-dependent and one part income-independent (the 'nominal premium'). The income-dependent premium was to be determined on an annual basis by the government at a level that fell short of the effective cost level. This premium (collected by the revenue department of central government) was to flow into a national fund (the Central Fund) from where it would be divided between the insurers concerned, thus helping to cover their expenditure. As this money would not cover all operating costs, the insurers would need to raise additional money by means of a nominal premium. Allowing insurers the freedom to determine the level of this premium would create a competitive arena, in which they would be able to undercut their competitors.

To ensure that the mandatory and voluntary schemes were competitive, it was necessary to change their legal position. So proposals were put forward to eliminate legal rules that prevented sick funds from (1) working outside their local region, (2) offering indemnity insurance, (3) denying contracts to licensed health care providers, or (4) undercutting officially fixed tariffs.

The Dekker Report aimed to introduce entrepreneurial risk in the social health insurance industry. The proposals were designed to end a situation in which social health insurance managers roll over operating deficits to either the government (by asking for additional public funding) or to the insured (by upward adjustment of premiums). The report wished to create a situation in which sick funds and indemnity insurers alike would develop client-oriented market strategies, thus preventing their clients from switching to cheaper and/or qualitatively better insurance alternatives. It was assumed that (nominal premium) price competition would be a major element in these market strategies, forcing health insurance managers to become proactive in terms of cost containment (van de Ven 1987). Therefore, the report provided insurers with instruments for cost containment: freedom of contracting with providers (replacing a licensing system that entitles every licensed organization to claim a budget) and replacement of (national) fixed tariffs by maximum tariffs that could be undercut in contract negotiations. In short, the proposals were intended to end the situation in which it is easier for social health insurance managers to pass on their operating deficits to the government and/or their clients than to undertake strategic initiatives to prevent such deficits by actively managing the expenditure of health providers.

In terms of solidarity, the Dekker Report can be seen as proposing another 'half-way' arrangement. There is full solidarity in terms of mandatory insurance rights for the whole population, thus ending differences according to socio-economic status. However, the premium to be levied is not fully income-dependent. The nominal premium, by its very nature, has disproportionately negative effects on the level of consumption of low-income groups. The Dekker Report does not end this disproportionality.

As of the end of 1994, these proposals had been partially implemented. The Central Fund for income-dependent premiums had been established and the nominal premium had been introduced. Legal rules establishing regional monopolies for sick funds had been withdrawn, creating direct competition for clients between the funds. And regulations have been introduced that gradually will end the obligation of insurers to offer contracts to all licensed health care providers (van de Ven 1993). The funding of health providers, based on license and budget negotiations, has yet to be replaced by contracting between insurers and providers, except in the areas of rehabilitation medicine and home nursing. The limited experience in these two areas suggests that in practice opportunities for insurers to cancel existing relationships and to switch to more efficient or cost-effective providers are limited – mainly because demand for acute and long term care in most areas is greater than that offered by providers.

The change to a two-part (mandatory/voluntary) insurance scheme has yet to be accomplished, due in part to controversies about the likely costs of such a change. Moreover, after the 1994 elections, all political parties announced that they had dropped the idea of a single comprehensive, mandatory social insurance scheme. The present Dutch Government supports a three-fold approach: mandatory insurance for long-term care, mandatory insurance for acute (ambulatory and hospital) care, and voluntary insurance (or direct payments) for everything else not covered by the first two schemes. Currently, differences between sick fund insurance and private (indemnity) insurance will be eliminated in a step-by-step approach. The government has opted for full financial solidarity in its long-term care insurance arrangements, by making the premium fully income-dependent. The acute care scheme will provide partial financial solidarity, by introducing a 200 florin (approximately US$120) direct annual payment for the use of health services ('own risk').

The decision regarding acute care is an attempt to try and end the chronic political debate on the introduction of direct payments and deductibles. The central controversy concerns income policy. Nobody denies that direct payments and deductibles have a disproportionately negative effect on the level of consumption of those on a low income. Supporters of the introduction of direct payments and deductibles argue that health insurance policy must not become blurred with income policy. So in individual cases, people need to be compensated for income effects using policy instruments outside the health insurance arrangements (e.g. by tax measures). Opponents argue that direct payments and deductibles will force low-income groups to under-utilize health services. This is considered bad policy, since the health status of such groups is already worse than others in the population. The government has sided with the supporters of direct payments and deductibles, and tax measures are promised to compensate low-income groups. The government has also decided to restrict cover for acute care in the areas of dentistry and physiotherapy. Whether this is an attack on solidarity depends on whether one views these particular health services as necessary or as luxuries (Commissie Keuzen in de Zorg 1991). As an unequivocal answer cannot be given, the debate regarding mandatory insurance for acute care will continue.

HEALTH INSURANCE MARKET STRATEGIES AND THEIR EFFECTS ON EQUALITY

The introduction of market forces in the Dutch social health insurance arrangements was only partially complete at the end of 1994. From 1987, sick fund insurers have taken the changes seriously; however, this does not mean that their views correspond with political intentions. From a political point of view, market forces were introduced to create incentives for cost containment. From the strategic point of view of health insurance managers, market forces imply a slightly different challenge: they introduce (at least in the long term) serious entrepreneurial risks that can threaten the survival of insurance organizations, so strategies have to be developed to secure continuity by minimizing operating risks (de Roo 1993).

The first strategic move was to scale up the size of the organization. Since 1987, the number of health insurers has fallen dramatically because of the ongoing process of mergers. This has created a lot of anxiety among health providers, who fear a weakening of their negotiating position in future contracting. At the moment, this fear has diminished. Health providers have realized that they have a strong negotiating position because demand for health services in general exceeds existing capacity. The health care market in this respect is a sellers' (providers') market rather than a buyers' (insurers') market.

Scaling up, however, is not enough to minimize the entrepreneurial risk for health insurers, and so additional strategies have been developed. The most important, the second strategy, aims to change the composition of the client group, leading to over-representation of 'good risks'. This can be accomplished by means of block insurance contracts with large organizations, for employees not covered by the sick fund. So in recent years, strong competition has developed between insurers in the block contract market sector. This competition has resulted in a levelling down of premiums, neutralizing the competitive advantage of creating an over-representation of good risks in the client group.

A third strategy is to disenrol bad risks. Opportunities for direct demarketing are small, due to legal rules that forbid insurers from excluding clients. The main opportunity is to encourage bad risks in the indemnity insurance market sector to switch to the Standard Insurance Scheme. As explained above, deficits in the operating costs of this 'half-way' social insurance scheme are covered in full by a supplementary premium, an annually adapted flat fee paid by all subscribers of indemnity insurance. In practice, this means that there is no entrepreneurial risk in this insurance. So shifting bad risks to the Standard Insurance Scheme is a worthwhile strategy. And it even can be argued that it is a strategy that contributes to greater financial solidarity, since the good risks in indemnity insurance cover the deficit in the operating costs of the Standard Insurance Scheme.

While these three strategies do not reflect the political goals of the change towards market forces, a fourth one does: health insurers have become more active in the area of cost containment. Activities are still restricted in range as health care providers' funding by budgets has been replaced by contracts in the fields of rehabilitation medicine and home nursing care only. But budgets have become tighter and insurers have become less willing to allocate additional funding. Providers of health care are expected to react to this by raising their productivity. Where providers see no opportunities for improvements in productivity, they shift the problem onto their clients by introducing waiting lists and by increasing the clinical criteria for care. This is happening in a growing number of fields: waiting lists are increasing for elective surgery and home nursing care, and the clinical criteria have changed for caring for the elderly and the handicapped.

Such barriers to access erode solidarity in two ways. First, indemnity insurers compete by offering their clients opportunities to avoid waiting lists for acute hospital care. This creates a widening gap between sick fund members and subscribers of indemnity insurers in terms of access. Such a gap also develops where hospitals have introduced 'company departments'. A recent change in Dutch social insurance legislation states that employers must pay the salary of a worker for the first six weeks that they are ill. This has introduced a strong incentive for companies to reintroduce their employees back into the workforce as soon as possible. Hospital 'company departments' exploit this situation by offering priority treatment to the employees of companies willing to pay for this service.

The growing interest in cost containment by insurers has had an additional side-effect that is eroding solidarity. Hospitals traditionally did not worry about the insurance status of their patients. In cases in which neither insurers nor patients could be billed (e.g. illegal immigrants, health care tourists, etc.), either the social security agencies picked up the bill or the hospital accepted a loss. It has now become normal practice for hospitals to secure payment by new patients, either by checking whether they are covered by health insurance or by requiring a deposit or guarantee. In 1994, such procedures were in place for non-acute hospital cases only. But this development illustrates the extent to which hospital managers have become more businesslike in their attitude towards patients.

THE EROSION OR AN UPDATE OF SOLIDARITY?

The foregoing illustrates the political search for solidarity in health care (in terms of cost-sharing and access) in an insurance-based system. An intrinsic lack of solidarity is shown by the difference between the mandatory sick fund and other health insurance schemes. This has generated comprehensive efforts to eliminate this difference, including the introduction of mandatory insurance for the population as a whole. The main drive behind these efforts is political ideology. At the level of daily political practice, the Dutch case shows a continuous search for incremental measures to improve the insurance arrangements for vulnerable (mainly low-income) social groups. This more practical approach is in line with the plans of the present government.

The search for a practical solution has given birth to 'half-way' insurance schemes. These schemes have now been brought together as the Standard Insurance Scheme, and it was explained above why indemnity insurers have an interest in enlarging their market share in this particular scheme. The Dekker Report and subsequent proposals can be understood as a long-term trend towards replacement of the sick fund and of indemnity insurance by a set of 'half-way' arrangements. However, Dutch politicians have demonstrated that they cannot find an objective way of distinguishing between necessary (e.g. mandatory) and luxury (e.g. voluntary) health services.

In practice, cost containment considerations have taken over, leading to political decisions to transfer what are in fact necessary health services to the voluntary insurance realm. The 1994 governmental plan to leave dental care to the voluntary insurance or direct payment sectors illustrates this line, thus eroding the equality of access dimension of solidarity. Cost containment overrides solidarity as a political goal. The long-term development towards a 'half-way' arrangement can also be seen in the growing political support for the introduction of direct payments. This support culminated in 1994 in the decision of the newly elected government to introduce an annual 200 florin direct payment for the use of acute health services. Here again cost containment has overridden solidarity as a political goal.

The 1994 government has announced plans to compensate low-income groups for the financial consequences of the loss of solidarity by means of tax measures. However, there is a striking difference between specific, operational governmental decisions on changes in health insurance and the non-operational promise to compensate low-income groups for the financial aspects of the erosion of solidarity. And no measures have been announced to counter growing differences in access. Such differences in access arise in the first place due to the reactions of health care managers to the cost containment activities of insurers. The government has also announced additional cutbacks in the growth of the macro-budget for acute care, increasing the gap that already exists between demand and supply. Under such conditions, it can be expected that managers will intensify their efforts to shift the cost containment problem onto the patient. The tendency towards growing differences among income groups in access to care may well be reinforced by these efforts.

Does this mean that we really are witnessing the end of solidarity in Dutch health care? Is there no more optimistic view about what is going on? To find an answer to this question, we have to take an even broader socio-historical perspective than that taken above. In the Netherlands, solidarity in insurance-based health care arrangements was institutionalized by the continued existence of coalition government, forcing four political parties to negotiate and accommodate their ideological views. Four political ideologies strongly influenced the efforts to operationalize the political concept of solidarity in health care policies (Banning 1964). The liberal interpretation of solidarity in Dutch health care was based on the thesis of equal individual human rights, as modified by J.S. Mill's view that governmental intervention was appropriate in order to protect individuals from evil caused by other individuals. The Dutch socialist interpretation of solidarity was influenced by J. Jaures' ideas about a cooperative ordering of social relations based on collectivization of resources and, in the 1930s, by the *plansocialist* H. de Man, who believed it necessary to use collective resources for social welfare goals. The view of the Calvinist parties was embedded in the political goal of creating a society in which 'people can live a life, carefree and worthy of a human being, according to the standards of their social status' (Banning 1964).

The Catholic Party had the most explicit and influential view on solidarity. Here the interpretation was not primarily derived from subjective ethics, from the feeling of love of one's neighbour (as in Calvinism) or feelings of individual (as in liberalism) or social (as in socialism) interdependency. In Roman Catholic social thinking, solidarity was an inherent aspect of organic social order, expressing the objective complementary and reciprocal relationships between capital and labour. Social order here meant collaboration between capital and labour under a functional norm of social justice.

The historical development of the Dutch health care system can be seen as a function of the political interaction between these four ideological frameworks. At the end of the nineteenth century, a political debate began about how to transform existing health care arrangements so that they were based on solidarity. Step-by-step, subsequent political actions led to the present health care system. This system can be understood as an institutionalization of operationalized political values of solidarity, as developed between the end of the nineteenth century and the first decade after the Second World War.

Judged by the political values derived from these four ideologies, the changes in the Dutch health insurance arrangements imply a net loss of solidarity. But how valid is this judgement? The political value of solidarity was operationalized under social conditions that were vastly different from today. Before the Second World War, the social position of workers was characterized by a low income, poor education, a poor health status, and bad working and housing conditions. A different perspective is required when viewing the changes of the last decade, and the way in which solidarity in health care insurance has been institutionalized. It could be argued that the changes do not reflect the fall of an important social value. There is evidence to suggest that we are witnessing an ongoing process of pragmatic socio-political adaptation of a political value to new social conditions. From this perspective, the changes described above can be understood as a process of updating the interpretation of an important political value.

REFERENCES

Banning, W. (1964). *Hedendaagse Sociale Bewegingen*. Arnhem: Lochum Slaterus.

Commissie Keuzen in de Zorg (1991). *Kiezen en Delen*. Den Haag: DOP.

Commissie Structuur en Financiering Gezondheidszorg (1987). *Bereidheid tot Verandering*. Den Haag: DOP.

de Roo, A.A. (1993). *De Zorgsector als Bedrijfstak in Wording*. Tilburg: Tilburg University Press.

Maarse, J.A.M. and Mur-Veerman, I.M. (1990). *Beleid en Beheer in de Gezondheidszorg*. Assen: van Gorcum.

van der Hoeven, H.C. and van der Hoeven, E.W. (1993). *Om Welzijn of Winst*. Den Haag: Azivo.

van de Ven, W.P.M.M. (1987). 'The key role of health insurance in a cost-effective health care system: Towards regulated competition in the Dutch health care market'. *Health Policy 7*, 253–72.

van de Ven, W.P.M.M. (1993). 'Health care reforms in the Netherlands: A demonstration project for other countries'. In D.P. Chinitz and M.A.

Cohen (eds), *The Changing Role of Government and the Market in Health Care Systems*. Jerusalem: Brookdale Institute.

van Mierlo, J.G.A. (1991). *Particulier initiatief in de Gezondheidszorg*. Assen: van Gorcum.

3

REGULATION OF PLANNED MARKETS IN HEALTH CARE

Göran Arvidsson

INTRODUCTION

This chapter deals with the regulation of planned health care markets. Its primary aims are to conceptualize: (1) what regulation is, what can be regulated, why, and by whom; (2) what a 'planned market' is and what kind of regulation is specific to a planned market, compared with a planned hierarchical system and a regular 'unplanned' market; and (3) what is the meaning and function of regulation in a planned market for *health care*.

The framework within which health care is provided contains many important elements that are not treated in the chapter, including demography and family structure, socio-economic factors, political institutions, taxation and social security systems, and the characteristics of the medical professions. Each health care system is a part of a much larger societal system. Although I am most familiar with the Nordic health care systems, in which elected public officials exercise direct public control over both the finance and operation of nearly all health care providers, in the analysis below I do not limit the discussion to any particular national setting.

THE CONCEPT OF REGULATION

As a point of departure, a rather general definition of regulation will suffice. Regulation will be understood as (1) the rules (norms and restrictions), (2) issued by (or on the behalf of) a government body, (3) aimed at influencing activities, (4) undertaken by actors within a defined society (or sector of society). The first point distinguishes regulation from other forms of influence, such as incentives, persuasion and *ad hoc* interventions. The second point excludes rules which are mutually agreed on by autonomous parties, for example as in contracts between independent purchasers and providers. The third and fourth points need not be elaborated upon further.

A basic concern of this chapter is the reasons for government regulation. There are at least three, each of which is related to a certain type of regulation. First, one function of regulation is to stabilize systems which are in principle self-regulated (e.g. families and commercial markets). Such regulation, enacted and enforced by legislation, means to create 'rules of the game' and codes of behaviour. The second type of regulation is used for sanctioning direct intervention in such systems, for example price controls. The third type is used for controlling the use of public resources (e.g. public investment and consumption, or for transferring payments to private entities). All these have – or are supposed to have – a specific function.

Regulation should not be confused with legislation. Regulation generally takes the form of, or has its legitimacy based in, legislation. However, legislation may also be used for symbolic reasons, for example influencing morale and attitudes. This type of legislation is not intended to be legally enforced. Also, legislation expressed in the form of laws may be implemented by subordinate regulatory measures (directives, decrees, ordinances, etc.). Our concern here is with the function of regulation, not the legal forms it may take.

Government regulation of a market is superimposed on the actors in the market by the state. It is vertical. An alternative is self-regulation, by which the actors themselves impose restrictions on their actions. Self-regulation may be initiated and organized by, for example, an industry association in order to avoid government interference. It is lateral in the sense that it is self-imposed. This does not necessarily mean that the rules set by self-regulation are less binding or are less of a hindrance. It is important whether the rules set by self-regulation are mandatory or not and how and to what extent they can be influenced. For individual actors, it may be more difficult to contest rules which are set by self-regulation than those which are based in legislation.

A more narrow and precise definition would be to restrict the concept of regulation to those *rules* which (1) restrict the actions of existing or potential actors, (2) are applicable under specified, predetermined circumstances, and (3) can be enforced by legal sanctions. Within this narrow definition, goal statements would not be included, even if they were legislated for. Direct interventions in specific cases would also be excluded, as would the use of administrative powers. They represent other categories of control instruments. Point (3) would exclude regulation by private institutions, as long as these institutions were not legally commissioned to impose sanctions. Such a narrow definition would, therefore, exclude a wide range of activities usually understood as regulation. For the sake of relevance, I will therefore use the wider concept of regulation as defined above.

PLANNED MARKETS FOR HEALTH CARE?

The supply, organization, funding and management of health care is on the agenda of the public policy debate, both nationally and internationally. Pressures from an ageing population and expensive new technologies, together with little improvement in productivity, are causing financial concerns. The inequality of access to care is a burning issue in some countries. As a consequence, deciding the criteria for rationing and prioritization is no longer the sole responsibility of the medical profession at the clinical level. The Oregon list in the USA – now endorsed by the Clinton Administration – and the proposals of government committees in New Zealand (Core Services Committee 1992), the Netherlands (Report of the Government Committee on Choices in Health Care 1992) and Norway (NOU 1987: 23) are just a few examples. In Sweden, such matters are currently on the agenda of the county councils, the bodies responsible for most health care provision. At the national level, a government committee is presently reviewing criteria for prioritization.

In each country, the same questions are pertinent: Who should decide what? What should be the role and responsibilities of government bodies (politicians and government officials), funding institutions, health care providers, individual clinicians, patients? It is evident that the answers to these questions are, to a large extent, dependent upon the institutional framework within which each individual health care system operates.

Dissatisfaction with the present systems has led to major reforms in recent years in some countries, including the UK. In other countries, the strategies for reform have been more cautious and the process much slower. In the USA, where market dynamics create constant adjustments of supply and demand, proposals by the Clinton Administration and others for major health reforms have an uncertain future. In Sweden, a government committee is presently reformulating how health care and the medical services should be funded and organized (HSU 2000). There are several options. A shift to an insurance-based system with competing providers of care as demanded by the neo-liberals, however, seems less likely than a facelift for the present, tax-funded and public sector operated system.

The following are indicative of the trends in the search for better health care systems:

- In countries where health care is to a large extent the responsibility of government, the trend is to introduce market solutions, with or without a relaxation of public funding. In market-based systems, the proponents for change tend to argue for more unitary systems and a greater degree of government intervention.
- In countries with a high degree of tax funding, the searchlight is on other countries' insurance-based solutions. In countries where most of the finance for health care passes through parallel insurance plans, there is a search for more coordinated and comprehensive systems.
- In centrally planned systems with predominantly public provision of services, the preference is for market-type solutions: a purchaser–provider split, internal (and perhaps some external) competition, profit centres, patient choice, etc. The traditional emphasis upon improvements in public health is gradually being replaced by strategies promoting *individual* health. In market-oriented systems, however, the change seems to be in the opposite direction; the problems of fragmentation, unequal coverage, lack of total cost control, etc., have spurred an interest in 'managed care' and *public* health strategies.

These opposite trends have raised the question of whether we are witnessing a convergence of the various systems. From a normative point of view, the question is whether a mixed model is possible and preferable. The concept of 'planned markets' (Saltman and von Otter 1992) would appear to build on such a belief. It is tempting to reject this idea immediately by pointing to the failures of the planned markets and administered price systems of Eastern

Europe, perhaps most clearly formulated and implemented in Hungary some decades ago. The Scandinavian 'mixed-economy model' is not an accepted ideal either. On the other hand, as Kornai and other have pointed out in their analyses of economic systems, 'To ask, "Planning or market" – is to ask the wrong question. Rather, what we must deal with are two, *complementary* control subsystems of the complicated and complex economic system' (Kornai 1971: 334).

However, health care constitutes just one sector of the economy. There is nothing to say that all sectors in an economy should be designed in the same way or have the same degree of government intervention. Health care should therefore be analysed independently.

In this chapter, the focus is on *planned* markets for health care. What is specific for a planned market, and what is the role of regulation in such a context?

PLANNED MARKETS

'Planned markets' are just one of a number of (at least theoretically) possible alternatives of resource allocation and coordination in health care. Three basic models – or principles – can be distinguished: market (buyer–seller interaction), hierarchy (planning and command chains) and solidarity (voluntary contributions and cooperation) (cf. Kaufman *et al.* 1986).

In the *market model*, the provision and consumption of health services, as well as research and development (R&D) and investments, are the result of numerous decisions and actions by independent (market) actors. Coordination is achieved by the 'invisible hand' of the market. In 'pure' health care markets, there are no central authorities planning the provision of health care; that is left to the interplay of consumers and providers. The guiding principle for the providers – at least in the long term – is economic profit. Patients are guided by effect–cost or benefit–cost ratios. Only those providers able to generate enough revenue to cover their costs will survive. The restrictions and rules governing the actors in a market model would be similar to those operating in other markets (e.g. market legislation and safety legislation).

In the *hierarchical model*, a governing body has overall responsibility. The provision of health care is coordinated, which means that providers are not independent. In its simplest form, a central body plans and directs the activities of local operating units. The scope and degree of local decision-making is a matter of delegation. The relationship between the central body and the local units – and between the local units themselves – may take many different forms, from a strictly regulated command chain to more or less market-like relations. The armed forces and public school systems offer good, but in many respects, very different examples of the hierarchical model. The former planned economies in Eastern Europe and traditional business firms are other examples.

In countries in which health care is considered the responsibility of government, the preferred organizational principle is one of hierarchy. The concept of a goal-oriented health care *system* implies a set of common goals, an organization and organized activities to fulfil these goals. Implementation by a hierarchy means that the producing units have their superiors, and that their activities are – or can be – legitimately steered and controlled from above. This is very different to a market, in which the actors enjoy autonomy within a general set of rules. Consequently the implementation of a national policy will rely on different instruments in a market model compared with a hierarchical model (cf. Williamson 1975, 1986; Lane 1985).

In the *solidarity model*, resources are allocated and activities are undertaken according to a commitment to 'do good'. Today's health care institutions are rooted in yesterday's charities (religious or not), workers' unions, and family and community networks. Voluntary health care organizations based on the principle of solidarity still play an important role at the micro-level; that is, as providers of health care within a 'market' or 'hierarchical' system.

For well-documented reasons, the 'pure' market model is rejected almost universally in the case of health care. Technical terms such as 'information asymmetry', 'moral hazard', 'adverse selection', 'externalities' and 'benefits of scope and scale' express some of the motives. Government interventions, therefore, restrict or complement market activities. The public funding and public provision of a health care infrastructure and services complement or substitute for private solutions. In market-oriented health care systems, there is a growing interest in institutional arrangements for facilitating coordination and cost control. 'Managed care' is one example.

For equally well-known reasons, the hierarchical model (in the form of a comprehensive, government supervised, funded and operated health care system) seems to have passed its peak, both in theory and in practice. Sub-markets for certain health care products and services are being created or restored to varying degrees in different countries. Market-like mechanisms, such as internal markets and sub-contracting, are being introduced in several government-operated systems. The British National Health Service (NHS) (see Bartlett 1991) and the Stockholm County Council regional health system (see Anell and Svarvar 1993) are just two examples. Usually, competition between providers of care is promoted, while financing is still kept under central control.

The serious shortcomings of both the market system ('market failures') and the planned, hierarchical government system ('policy failures') are a reason for the ever-changing financial and institutional forms of health care provision. In some countries, the changes mean more market and less planning; in others, it is the reverse. The trend seems to be to search for the optimal combination of market and hierarchy, of the role and influence of the invisible and the visible hand.

The concept of a 'planned market for health care' is just one of several possible combinations. As I understand it, this model differs completely from a 'regulated market' in one crucial respect: the existence of a *principal*. A *planned* market implies that there is someone that plans for the system, that there is someone in charge and with the authority to guide the activities. A planned market implies the existence of a principal. (The principal may be the *agent* of a collective body of patients or taxpayers.)

In a *regulated* market, there are indeed rules and regulations restricting the actions of individuals and organizations, but there is no higher authority informing them what objectives to pursue and what activities to undertake. The markets for food and clothing in western countries are examples of regulated markets. Legislation concerning forms of competition, consumer safety, trade barriers, taxes, etc., may have a significant influence on both producers and consumers, but there is no authority charged with optimizing the production and delivery of food and clothing, and there is no central responsibility for cost containment in the food or clothing delivery systems. This is taken care of by market forces. There is no principal.

In the case of health care, government policy usually states the overall objectives. And because public money is involved, there is a need for resource allocation and financial control processes in the execution phase. In these respects, there are no differences between planned and regulated markets for health care. The crucial question concerns the government's authority and accountability. In a planned market, government can legitimately intervene by

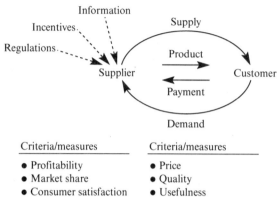

Figure 3.1   The market

fixing prices, setting production targets or making structural changes (e.g. it can open and close hospitals). These are examples of the responsibilities of a principal. In the case of a regulated market, government responsibility concerns the operation of the market, or rather the outcome of market operations. In addition to laying down the rules for market operators, government may act as an operator itself – that is, as a buyer or a seller. Examples may be found in financial markets, for example.

Regulated and planned markets may co-exist even within the same sector of society. In health care, national government may act as a market regulator, whereas regional governments may assume the role of the principal of regional, planned markets. This is the model which is being implemented in Sweden, a country which used to have a more or less hierarchical model (with most responsibility lying at the regional level).

Government intervention concerning the licensing of doctors and other professionals, setting medical standards, medical supervision, etc., are examples of control mechanisms necessary in all types of system – competitive markets, regulated markets, planned markets and government-operated systems – and should not be confused with the function of a principal. In fact, responsibility for professional standards and ethics is often 'delegated' to – or voluntarily assumed by – professional bodies, more or less formally linked with government.

In general analyses of most markets, two categories of actors can be distinguished – producers and consumers – the relations between whom are clear-cut. The producer delivers a product and gets paid. The revenue is, in turn, used for compensating the producer's employees, suppliers, etc. All these transactions take place within an institutional framework that is common to all actors of a similar kind (at least in principle). The decision criteria and performance measures are relatively straightforward (see Fig. 3.1).

Descriptions of health care markets usually include a third category of actor – that is, funders/third-party payers – the reason being that some type of insurance is necessary. The result is the market triangle, in which the roles and interrelations of the three types of actor can assume many different forms. The existence of a principal is neglected or implicit. This is probably the key dilemma of a market-oriented health care system. Coordination and re-source allocation are not taken care of by the invisible hand of the market, and there is no visible hand to do the job either. In Fig. 3.2, the funder as well as the principal are included. It is assumed that the principal is a government body.

In tax-financed health care systems, the roles of the principal, the funder and the producer are typically integrated within the public sector. This minimizes transaction costs (which exist between par-ties in a market; see Bartlett 1991). Instead, there are the visible or invisible administrative costs of hierarchies.

Criteria/measures	Criteria/measures
● Diffuse or controversal ultimate goals	● Cost
● Non-monetary output measures	● Quality
● Indications	● Availability
	● Benefit

Figure 3.2 A market with a principal and a third-party payer

The models in Figs 3.1 and 3.2 are both at the micro-level. They illustrate one specific transaction and include the actors, transfers, rules, etc., which are involved. At the macro-level (e.g. a nation), there are of course many consumers and providers. The latter may be private or public, for-profit or not-for-profit, large or small, autonomous or part of a business group or government, etc. In Saltman and von Otter's (1987, 1989) model of 'public competition', the provider institutions are publicly owned and operated but managed in a businesslike manner. Consequently, they are called 'public firms' (von Otter 1991: 66). In principle, it is not necessary for providers in a planned market to be publicly owned and operated. It is sufficient that the providers, public or private, agree to act as agents of the principal. This means that they have a commitment not only to their customers (individual patients), but also to the principal (as the representative of a collective interest). In practice, this is probably most likely to be the case if the private providers are relatively few and receive a large proportion of their revenue from the principal, either directly or according to rules set by the principal (as in a general, mandatory health insurance system).

There is no self-evident relationship between a principal and a funder. The case of a principal also being a single-source payer is just one of many possibilities.

In the real world, there are many funders: patients, taxpayers (via government bodies) and premium-payers (via insurance institutions) in various mixes. The funders' relationship with the other actors must be considered in an analysis of a planned market. And to complicate the picture further, there may be many principals. In health care, it is not unusual to find a principal at each level of government: nation, region, municipality. The allocation of responsibilities and powers may take many forms, both vertically (authority and command chains) and horizontally (lateral networks between regional and local health care authorities).

In the following section, alternative models of resource allocation and control will briefly be examined. The purpose is to highlight the specifics of a planned market.

THREE MODELS OF RESOURCE ALLOCATION AND CONTROL

As soon as a principal has certain (enforceable) objectives which affect the activities of the providers, it would be more appropriate to talk about a goal-oriented (purposeful) system than about a market, in our case a health care *system* rather than a health care *market*. Now, let us consider a field of activity (e.g. health care) in which there is a principal and an unspecified number of producer units serving a more or less well-defined group of recipients (e.g. a local health care authority) and various health care providers (clinics, etc.) serving a population. Let us focus on the relations between the principal and the producer and, therefore, assume that there is no third-party payer in addition to the principal. The crucial question concerns what services to provide and to whom. This in turn raises the question of how production and investment decisions should be made and by whom.

Suppose that the central authority, the principal, wishes to optimize the total effectiveness of the system on behalf of the population. This can be achieved, at least in principle, by institutional arrangements or by direct intervention in the production process, or by a combination of the two. Suppose also that the principal can choose from various combinations of administrative (directives, regulations, etc.), informative (communication of knowledge, propaganda, etc.) and economic (fees, budgets, etc.) means to guide and control the production and delivery of services.

At least three different models can be distinguished with regard to the roles of the principal and the producer units in allocating resources and making production decisions. Using concepts related to the form of resource allocation, they may be labelled 'quantity allocation', 'budget allocation with resource pricing' and 'administered price allocation'. All three models are treated in the management control literature (albeit under different labels), and all three are easy to illustrate with examples from the health care sector. These will be considered in turn with the purpose of clarifying the specific characteristics of a planned market.

In the first model, *quantity allocation*, no market mechanisms are utilized. Resources are allocated to the producing units in physical terms; for example, doctors (positions or man-years), nurses, localities, facilities, etc. The production process is controlled by administrative and professional means (guidelines, manuals, protocols, inspections, etc.) and outputs are defined in physical terms (visits, consultations, examinations, operations, etc.). Depending on the type of activity, outputs may be targeted in advance or only estimated (determined by external demand). These targets and estimates are considered in the allocation of inputs. An iterative planning process seems reasonable. (We have to leave aside how

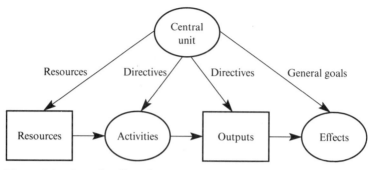

Figure 3.3   Quantity allocation

this could be done). Any payments from recipients are placed at the principal's disposal. Information concerning desired effects is communicated to the producers in the form of goal statements or priorities of a more or less general character (see Fig. 3.3). This is the 'bureaucratic command system'.

It is difficult to imagine this model being implemented at a national or regional level. One obvious reason is the vast amount of information that would have to be submitted to, and processed at, the centre. At the local level, however, this model would not be too unrealistic. Many hospital clinics and health care centres, not only in Sweden, are probably run along these lines – or at least they used to be before present-day financial control systems were introduced.

In the second model, *budget allocation with resource pricing* (Fig. 3.4), a budget in monetary terms is allocated to each producing unit. How this is done (i.e. the budgetary process by which the budgets are arrived at and decided) is, of course, of crucial importance for the overall allocative and distributive effectiveness of the system, but this will not be discussed here (cf. the budget and management control literature; e.g. Anthony and Young 1984). Decisions concerning the actual mix and quantities of resources are delegated to the producing units. As a decision parameter in these choices, prices are set for resources made available by the central authority. Resources may also be bought externally within the limits set by the budget. The production process, outputs and desired effects are controlled in the same manner as in the previous model.

This model may be described as a combination of a command system on the output side and a planned market on the input side. This is typical of the control of hospitals where the principal and not individual patients – or their insurers – are the buyers. Of the three

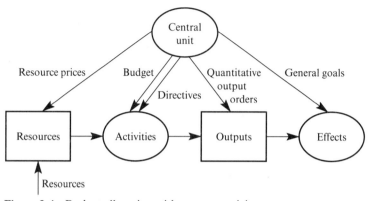

Figure 3.4 Budget allocation with resource pricing

models outlined here, this best represents the 'modern', tax-based, public health care system.

In the third model, *administered price allocation* (Fig. 3.5), prices are used to guide local decisions concerning inputs and outputs. This means that local producing units face markets for both inputs and outputs. They buy their resources and sell their products. In the 'pure' case, all the income of a producer would be represented by payments for services rendered. The goals of the producers are stated in economic terms rather than as quantitative production goals. As in the other models, regulations concerning resources and production processes may exist in order to ensure desired structural and process quality (Donabedian 1979).

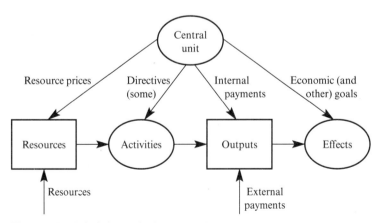

Figure 3.5 Administered price allocation

This is the case of a planned market for both inputs and outputs. It is clearly distinguishable from a normal market with autonomous sellers and buyers as well as from the quantitative and budgetary control models described above. It seems reasonable to reserve the term 'planned market' for this situation. More precisely, it is suggested that the following conditions should be met:

1   A principal is responsible for the overall objectives and priorities as well as for the overall production structure.
2   Providers of care have operational autonomy. They act as sellers and buyers within the limits set by institutional regulations.
3   Supply and production are influenced by the principal via incentives and guided by prices for inputs and outputs.

The last point deserves further comment. According to Saltman and von Otter (1992), a planned market is characterized by publicly planned and controlled supply. If this should be interpreted as the quantitative regulation of production capacities and outputs, it would require expanding the meaning of a 'market' beyond reason. Paying producers according to their output rather than by traditional budgets is not enough. It may be important from many respects, but a producer with no say over its supply and products is not operating in a market: it is a part of a hierarchy. In a planned market, at least some producers must be at arm's length from the principal.

As already noted, a *planned* market also differs from a *regulated* market. In a planned market, someone plans and monitors all activities and resources. In a regulated market, this is up to the market actors; the regulator sets the rules and acts as supervisor. The regulator's main instruments are rules (regulation!), sanctions if violations are committed and, perhaps, advice. In the case of a regulated health care market, it would mean that the governing body would leave to patients, third-party payers and providers to decide what services to provide, how and by whom. The task of the governing body would be to regulate access to the market, general contract conditions, standards, liabilities, etc. This governing body would not have the role of a principal. An insurance-based health care system, such as that in the USA, may be an example of a regulated market.

A regional or district system, as in Sweden and the UK, in which the regions/districts are responsible for public health promotion and for the provision of health care services, but rely on autonomous producers for the actual provision of services, may be an example of a planned market at the local level. At the national level, there are several alternatives. If national government wants to delegate all responsibilities for health care to local government, it may assume the role of regulator of the general conditions for local operations. It may also set national standards and evaluate outcomes. The national market for health care would be regulated, not planned. If, on the other hand, national government wants to enforce a national policy by being directly involved in the investment and production decisions, the result would be a planned national market. This means that a regulated market may include planned sub-markets. This does not facilitate the identification and analysis of different kinds of market regulation.

It is evident that Fig. 5 models a very simple form of a planned market. The model may be expanded in different ways to better represent reality. Additional actors may be included, such as third-party payers separate from the principal, and principals at several levels of government. Patients may, at least partially, be given the role of purchaser, and not just be the recipients of care.

Administered price allocation may take place in a 'closed' system or in an 'open' system. In the first case, all economic transactions take place internally. The resources needed for production are acquired from the principal or special internal suppliers. Final products are paid for by the principal, intermediary products by internal buyers. Prices are fixed or at least supervised by the principal. In an 'open' system, on the other hand, there is room also for external suppliers and buyers. (However, with a growing share of revenues coming directly to producers from external buyers, the situation would approach that of a normal market.)

REGULATION

What can politicians (reasonably) decide in a health care system (of the Swedish type)? The answer given recently in a speech by the health care director of one of the Swedish counties may be used to distinguish regulation from other policy instruments open to politicians. His answer encompassed:

- patients' rights;
- goals to be achieved;
- fair allocation of resources intended for purchasing care;

- principles of access to care;
- conditions/licences for access to taxpayers' money;
- global cost control;
- control systems and institutional forms of production.

In what he presented as his 'vision', there would be a balance between the national health authority, regional political bodies and producers (public, private for-profit and not-for-profit). The position of patients would be strengthened, but they would not have complete freedom to choose between providers and services. Purchasers of care would be staff-units of the regional political bodies. Their competences would include social medicine, health economics and knowledge of individuals' preferences. The remuneration system would be linked to production. Hospitals would have greater autonomy, but new building programmes, expansion, down-sizing and closures would be decided by a regional public body. Patients' rights to choose between providers was considered crucial, as was the existence of several providers, at least some of them independent of purchasers. Purchasers should be competent and powerful. His conclusion was that all this requires a much clearer set of rules.

I have cited this health care director's vision for two reasons: the system can easily be interpreted as a planned market, and it illustrates the place of regulation in such an environment. It is noteworthy, but not surprising, that greater freedom of the actors is considered to require new, and probably more, regulation. All seven points may be the object of legislation, but only for points 1, 4, 5 and 7 would regulation seem most appropriate. Points 1–4 may be implemented by informative means ('sermons') alone or in combination with regulation ('sticks') and incentives ('carrots'). All seven points may be implemented using economic incentives and disincentives, in combination with other means.

Since the various instruments are complementary, it would be misleading to analyse one without respect to the others. Being aware of this, I will nevertheless now focus on the potential use of regulation as a means to control a planned market for health care.

In order to structure the discussion, it would be helpful to use some kind of classification. One is to focus on the actors, the targets of regulation: purchasers (payers), providers and consumers (patients). From a managerial point of view, it would probably be illuminating to classify regulations in this way. It would help to distinguish between the rights and obligations of each party and to reveal inconsistencies. Besides, the monitoring of liability and accountability requires an identifiable agent.

Another alternative is to start by considering regulation as a means to create an institutional framework. In the present context, regulation could be understood to mean establishing and maintaining a planned market for health care. A next step could be to treat regulation of the supply and demand sides of that market. On the supply side, the following are examples of what could be regulated:

- entry/exit of providers (who, how many, etc.);
- conditions for establishment as a provider of services (criteria, accreditation, etc.);
- conditions concerning geographical location;
- forms of competition;
- standards of quality;
- prices;
- reporting requirements, including quality (technology, outcomes), costs, etc.;
- mandatory products to be offered;
- the authority to refuse to deliver/to discriminate between buyers and between patients.

On the demand side, regulation might concern the following:

- patients' rights (e.g. choosing between providers, treatments, non-medical services, etc.);
- conditions for establishing and working as a third-party payer (insurer);
- financing and reimbursement systems;
- the authority for third-party payers to discriminate between patients and between providers.

A brief review of the documentation concerning the Swedish health care system reveals that the regulation system is neither organized nor specified in terms such as these. The reason is probably that in a comprehensive, planned system (as the Swedish system has been and basically still is), this is not necessary. As a consequence of the recent, widespread purchaser–provider split and new, private providers entering the arena, there is bound to be a change, however.

A third alternative is to use a 'systems approach', which is probably more familiar to those individuals used to a planned system. Designing and operating a production and delivery system requires that decisions be made concerning (1) goals and strategies, (2) structure, (3) processes, (4) output/products and (5) outcomes/ effects. This offers a method for formulating a framework for regulation. For each of the five aspects, or dimensions, it is possible to imagine and consider different combinations of regulations, incentives and informative means. Examples of possible applications of regulation are given below. Several unclear issues arise concerning the specifics of regulating a planned market.

1 *Goals and strategies.* Overall objectives and priority rules may be legislated for. If, however, they are of a general, non-operational and non-binding character, this is control by information rather than by regulation. The Swedish Health Services Act, implemented in 1983, offers an example. It 'stipulates that the county councils shall plan their health care on the basis of the needs of the population' (Anell and Svarvar 1993: 14). This statement is interpreted and put into practice by the county councils; for example, by allocating resources to health care districts according to their perceived needs. This is done by a political process. It is not clear how such general goals could be communicated to and made to be respected by actors in a planned market. Stating acceptable 'corridors of activity' may be one way of translating goals and strategies into operative guidelines.
2 *Structure.* This is a promising field for a market regulator. The requisites for entry and exit of providers, competency requirements and staffing are obvious examples of what can be regulated. Regulation might take the form of authorization or licensing arrangements. But this creates a *regulated* market, not a *planned* market. Since the principal in a planned market can influence production by taking part in investment and operational decisions – as an owner of facilities and a buyer of services – the need for structural *regulation* seems to be less important. On the other hand, if the intention is to create and maintain a planned *market*, it is necessary to provide room for independent actors – and they need stable conditions (e.g. a commitment from the principal that they may belong to the 'structure' as long as they meet the competency and other requirements and perform adequately).
3 *Processes.* Process regulation seeks to control the production and delivery processes. Procedural rules may, for example, be used to set quality standards. In both competitive and regulated markets, it must be questioned whether direct process control is necessary

and desirable. Such markets are typically characterized by buyers and sellers acting as mutual controllers. In both cases, government has to offer institutions for settling complaints and disputes. Motives for organizing a planned market could be the nature of the outputs (collective or hard to define and evaluate), or the existence of uninformed or otherwise incapable customers/ recipients. In such cases, it may be considered necessary to control quality and service distribution by procedural standards, monitoring and evaluation. This is perhaps even the essence of a planned market. As an alternative to direct process control, contractual arrangements concerning quality assurance, etc., may be preferred. They must, of course, be backed by a sanction system that has been legislated for. Cases of clear malpractice may best be referred to a national authority.

4 *Outputs/products.* In a planned market, outputs would not be regulated but determined by market processes. Regulation may, however, be used to prevent discrimination against certain patients and other well-known market failures. Regulation in this area by national government would interfere with the authority of principals at the regional level. Regulation by county councils would, similarly, interfere with the authority of local principals or purchasers. Regulation may, however, enhance the rights of a patient at the expense of a principal/third-party payer. It is a delicate task to balance the interests of a principal, representing primarily a collective interest, and individual patient. If there is a powerful regional principal controlling the planned market, there may be a need for national regulation defending individual interests, as in the case of social services. (On the other hand, this immediately raises the question of who should pay.)

5 *Outcome/effects.* The purpose of regulation in this area would be to control the final effects of health care activities. This appears difficult if not impossible to achieve.

CONCLUSION

One obvious conclusion is the need to be much more precise about the meaning of 'regulation', especially if the purpose is to contribute to the design of a regulatory framework for a planned market. If all the possible ways of influencing performance by legislation and other formal, administrative means are considered to be regulation, it is difficult to go beyond a general statement that a principal has a wide variety of options, including that of abdication. Such a conclusion would not be very helpful.

More precise concepts, and a better knowledge of the character of each type of intervention open to 'market-makers' and 'market operators', would facilitate more qualified analyses of present and proposed health care models in a number of countries. It would also be much easier to come to a genuine understanding of the role of regulation in the context of a planned market, compared with, for instance, a regulated market system (like that in the USA or the Netherlands) or a planned system (like the traditional Swedish and British models). More research, both theoretical and empirical, is required. Prescription without an accurate diagnosis could be fatal.

REFERENCES

Anell, A. and Svarvar, P. (1993). *Reformed County Council Model*. IHE Working Paper 1993: 3. Lund: IHE.

Anthony, R.M. and Young, D.W. (1984). *Management Control in Nonprofit Organizations*, 3rd edn. Homewood, IL: Irwin.

Bartlett, W. (1991). 'Quasi-markets and contracts: a markets and hierarchies perspective on NHS reform'. *Public Money and Management*, Autumn, pp. 53–61.

Donabedian, A. (1979). 'The quality of medical care: A concept in search of a definition'. *Journal of Family Practice*, 9, 277–84.

Kaufman, F.X., Majone, G. and Ostrom, V. (eds) (1986). *Guidance, Control, and Evaluation in the Public Sector*. Berlin: Walter de Gruyter.

Kornai, J. (1971). *Anti-equilibrium*. Amsterdam: North-Holland.

Lane, J.-E. (ed.) (1985). *State and Market: The Politics of the Public and the Private*. London: Sage Publications.

Ministry of Welfare, Health and Cultural Affairs (1992). *Report of the Government Committee on Choices in Health Care*. Rijswick: Netherlands.

New Zealand National Advisory Committee on Core Health and Disability Support (1992). *Core Health and Disability Support Services Project*. Wellington, Department of Health.

NOU (1987). *Retningelinjer for prioriteringer innen helsetjeneste*. Oslo: Socialdepartementet.

Saltman, R. and von Otter, C. (1987). 'Re-vitalizing public health care systems: A proposal for public competition in Sweden'. *Health Policy*, 7, 21–40.

Saltman, R. and von Otter, C. (1989). 'Public competition *vs* mixed markets: An analytic comparison'. *Health Policy*, 11, 43–55.

Saltman, R. and von Otter, C. (1992). *Planned Markets and Public Competition*. Buckingham: Open University Press.

von Otter, C. (1991). 'Politisk styrning och marknadskrafter i sjukvården'. In G. Arvidsson and B. Jönssen (eds), *Valfrihet och konkurrens i sjukvården*. Stockholm: SNS.

Williamson, O.E. (1975). *Markets and Hierarchies: Analysis and Antitrust Implications*. New York: Free Press.

Williamson, O.E. (1986). *Economic Organization: Firms, Markets and Policy Control*. Brighton: Wheatsheaf.

4

CONTRACTING AND POLITICAL BOARDS IN PLANNED MARKETS
Mats Brommels

INTRODUCTION

'Planned markets' or 'managed competition' are two terms used to characterize ongoing changes in practically all Western health care systems. There appears to be something of a 'convergence of ideologies', a prediction from the late 1960s which suggested the emergence of a middle ground between open markets and planned economies. Health care seems to be one of the few sectors in which this scenario has in fact materialized. Previous US governments have introduced increasing regulation of the health care market and the Clinton Administration proposed a largely publicly controlled financing system. The public systems in Europe – both tax-funded and insurance-based – have been introducing competition into the provider side of their health systems. This development of convergence makes comparative studies across borders increasingly useful and also easier to accomplish.

The five Nordic countries – Denmark, Finland, Iceland, Norway and Sweden – form a culturally uniform region. They have a common history, the same type of societal system, a comparable standard of living and relatively good cultural and political cooperation. Health care is, by and large, organized in a similar manner. Nordic health care is of special interest from an international perspective in that it seeks to integrate primary, secondary and tertiary health care as well as social services and social benefit schemes within the same public service framework. As a public service, it is controlled by politicians on different levels, and their authority also extends to decision-making at an operational level. Thus two distinctive features of Nordic health care are (1) its comprehensiveness and (2) its political accountability. Nordic experience might consequently be useful to those countries where there is a lack of coverage of health care services and a perceived low responsiveness of service providers to consumers and citizens.

The overriding concern in most ongoing health care reforms has been to contain health care costs. In the public systems, the objective has been to increase productivity and efficiency of services, while maintaining the overall political goals of equality and improvement of the health of the population. The 'generic' solution – regardless of system characteristics or fine-tuning of the reforms – has been to introduce competition between publicly owned providers within a publicly funded and controlled 'market'. Similar arrangements to the British reforms – separating the 'commissioning' of services from provision, known as the 'purchaser–provider split' – are being introduced in other public systems. In this chapter, the British terminology will be used to describe these new organizational rearrangements.

In looking for differentiating characteristics among the tax-funded public systems (the UK and the Nordic countries), one specific Nordic trademark is the involvement of local *elected* politicians in decision-making. The British National Health Service (NHS), in contrast, is a national government agency, directly accountable to the Secretary of State. Executive and non-executive Board members on different levels are appointed by the government or its administrative offices. Thus the NHS is a public health care system with strong centralized control and funding. The Nordic systems are based on a decentralized service, with the responsibility for service provision resting with local authorities. Funding is shared between the periphery and the centre. Central control tends to be directly proportional to the central government's share of funding (Maxwell 1981).

When looking at these systems in more detail, interesting differences within the Nordic family of countries can be demonstrated (Brogren and Brommels 1990). In Sweden and Denmark, the health care services are run by counties (regional authorities) and in Finland by municipalities (local authorities). In Norway, the counties provide hospital and specialized services, whereas the municipalities are responsible for primary care. In Iceland, health care was recently centralized, with the government taking over the service, although the municipalities appoint some of the members of local boards.

The aims of this chapter are to explore (1) the role of politicians in initiating the health care reforms, (2) their influence on implementation and (3) their involvement in the new system of management control – that is, contracting – which was introduced as a cornerstone of the reforms. The third task focuses on the involvement of politicians in commissioning boards. The chapter tries specifically to trace any change in the role of politicians, and to seek reasons for differences between observed intentions and reality.

MATERIALS AND RESEARCH METHOD

The analysis will focus on the differences between the Swedish and Finnish health care systems. The main 'independent variable' is the level at which elected politicians are involved; in Sweden this is at the regional level, in Finland at the local level. In order to compare like with like, the 'observation unit' in Finland will be the province, which comprises several municipalities. The municipalities have established municipal federations within each province to run the secondary and tertiary care services. Environmental differences will be explored by looking at the origin of the reforms in both countries, and how they have been implemented.

The implementation of the reforms and commissioning in relation to the level of involvement and role of politicians will be considered in detail by systematically comparing five actual cases. Information about these cases has been collected from official documents, published papers and consultancy reports, and field study observations. The field studies cover all five cases, two being consultancies and three based on access to information in an internationally based information-sharing network.

The case studies cover three provinces in Finland and two counties in Sweden. In Finland they include a rural province with one municipal federation providing secondary and tertiary care (access by consultancy: FI-1), an urban province with three municipal federations providing secondary and tertiary care (access by network: FI-2) and a rural self-governing province providing secondary care (access by consultancy: FI-3); in Sweden they comprise a rural county providing primary and secondary care (access by network:

S-1) and an urban county providing primary, secondary and tertiary care (access by network: S-2).

THE REFORMS

Local self-government has a long tradition in the Nordic countries. Its legal basis, including its taxation rights, was gradually formed during the nineteenth century, and its responsibility for providing public services continued to increase until recently. National governments exercised control by legislative measures and policy direction. As an example, national health care licensing and accrediting authorities were professional bodies which gave instructions and advice to local authorities and institutions. Government also used subsidies to local authorities to help equalize differences in economic conditions, and to create incentives to establish or expand services. Since the 1960s, the public sector has undergone decentralization. In Swedish health care, this reached its height with the 1983 Health Services Act, which stated that the counties had overall responsibility for providing health care to their populations 'on equal grounds', but the counties were left to decide in which way and on what level. As a result, twenty-six distinct health care 'systems' have evolved, with differences in service mix and internal organization. The differences are to be found in the details, however, since the legal basis of county self-government and responsibility for arranging health care services is homogeneous throughout the country.

The most important features common to all Swedish county health organizations are a comprehensive service covering primary care, dental care and secondary specialized services (including hospital care), owned, funded and run by the county, with the decision-making executed by an elected county council and appointed health boards. Social services and housing, together with environmental health protection, are the responsibility of local authorities (municipalities). Sickness and disability benefits are administered by a government-controlled social insurance institution.

Finland lacks the regional administrative tier. The provision of social services, health care and environmental health protection are the responsibility of the municipalities. Municipal federations were established to provide specialized hospital care. These are independent organizations with a legal base in the Municipal Self-governing

Act. The 'member municipalities' own (i.e. pay for capital outlays) the federation and its facilities, cover their share of its budget (in relation to service utilization) and are represented on its council and board through appointed members. The locus of power is the municipal councils and health boards. Until 1 January 1993, however, the government exercised direct control over the municipalities and municipal federations through a strict central planning system, tied to government subsidies paid as a fixed percentage of actual costs. Consequently, the government and its provincial agencies approved and controlled expenditure quite closely.

Paradoxically, although Finnish health care has been controlled centrally, service provision itself is 'more decentralized' (i.e. to the municipal level) than in Sweden with its (regional) counties. The differences in organizational set-up have meant that reforms have been initiated and implemented with different objectives and with different timing.

The Swedish reforms stem from the rapidly deteriorating financial status of the counties and general discontent with the size and performance of the public sector. They were superimposed on the wave of decentralization commenced in the 1960s, aiming to both strengthen local decision-making and to involve the local population, and to increase efficiency and service responsiveness within the public sector (the front-line argument or the inverted pyramid organization). The delegation of economic responsibility and decision-making to hospitals and health centres, and further to clinical departments and 'basic units', was driven by the hypothesis that decisions should be made by those who 'were close to the customer, and knew the activity'. The main instruments were delegated authority and 'frame budgets' (Håkansson 1986), while the institutional structure and internal organization of the counties, by and large, were untouched or altered superficially. This development, which took place in the 1970s and 1980s, was widely termed 'decentralization', and resembles the *management by objectives* control model. The development was usually seen as positive and 'empowering' (Gabrieli *et al.* 1990), but there was increasing criticism during the late 1980s. Rombach (1991), for instance, noted the genuine differences in tasks, institutional framework and decision-making environment between a public administration and an enterprise.

The reform debate was also influenced by the strong neo-liberal political views of the early 1980s, the fall of the Soviet Bloc and the subsequent disdain for state-run economies. As in the UK, politicians and administrators in Sweden turned to 'competition' as a means to improve efficiency in resource utilization. Units were to 'earn' their resources. Budgetary allowances were to be replaced by performance-related income. 'Free utilities' within counties or hospitals were to be abolished, and cross-charges between departments were introduced. The discussions were basically about creating 'internal markets', as in the UK. Little dispute about the basic societal role of health care and other welfare services emerged. The reform agenda has only minimally been ideologically driven. Changes have been introduced in different counties regardless of political majority. Only the conservatives have actively promoted privatization; their coalition partners in the centre do not support the initiatives, however.

In the following analysis, only the organizational issue – the purchaser–provider split – will be pursued. The other major issues in the recent Swedish debate (i.e. patient choice, the family doctor initiative, and the government's review of health care funding) do not have the same relevance for this analysis.

What is of importance is that the *role of politicians* was made an important element of the reforms. Extensive discussions developed within the counties and especially among county and hospital administrators about the *real* task of politicians, and how that could be best *organized*. The prescription was: politicians should set the agenda, decide overall policies and strategies, be the true representatives of consumers, and stop interfering in operational management. The purchaser–provider split seemed to be the obvious solution. The place of the politicians was within the *purchaser organization.*

The indirect role of national government in Swedish health care meant that most of the reforms were initiated directly within the counties. Government agencies have been taking part in the national debate, however. The coordination of the debate has in the main been handled by the Federation of County Councils, an interest and pressure organization which in part substitutes for the government as a central policy-making body (cf. Crossroads Review 1991). The reforms are characteristically county-specific, with every county designing a 'model' of its own. In summary, the Swedish reforms were initiated within the counties, were focused primarily on health care, were driven by the need to improve efficiency in resource utilization, were built on an internal wave of decentralization, were carried out as organizational arrangements within the county, and included an intensive discussion on the role of political decision-makers.

The Finnish reforms and the resulting changes in health care organizational are also part of a major decentralization thrust within the public sector. Although the provision of most welfare services has been the responsibility of the municipalities, their activities have been strictly controlled by the national government through detailed legislation and financial control. The municipalities pressed for greater freedom, and the government became increasingly concerned about the inappropriate spending incentives in its subsidy scheme. Government subsidies were originally designed to encourage the municipalities to expand services (and thus the subsidies were activity-related, open-ended and covered a fixed percentage of actual costs). Even when a sufficient level of services was provided, the municipalities still had the incentive to improve further (e.g. in order to create new government-subsidized jobs). The government subsidies turned into the fastest growing part of the national budget, and have since the 1970s been a constant source of concern to the Department of Finance. It has pressed for reforms of the subsidy system throughout this period. Resistance to change was strong, both on the part of the municipalities who were afraid of losing out under any reforms, and on the part of government sector agencies and their administrators, including the National Board of Health of the Department of Social Welfare and Health. In many cases, members of parliament are heavily involved in local politics, and the municipalities have strong support in government decision-making across party boundaries. Under any reform, the sector agencies were worried that they would lose their influence and control over service provision, tied to the subsidy and planning system, which required plans to be submitted in advance and expenditure to be checked in detail *ex post facto*. The biggest government-subsidized sectors are (1) social services and health care and (2) education (primary and secondary schools and vocational training).

The government subsidy system was reshaped in three phases. Beginning in 1984, subsidies to the social and health care services were paid to the municipalities according to the same criteria and percentage of expenditure, correcting the prior bias in favour of health care. This reform was seen by the sector authorities as a shift towards the use of social support and community services in place of hospital care. In 1988, 'minor subsidy items' for a wide range of services were lumped together, and paid to the municipalities on the basis of the size of their population. Finally, in 1989, the Permanent Secretary of the Department of Finance, Teemu Hiltunen, was asked to propose plans for the total reform of the subsidy system.

His plan – basically a shift from the old activity-based, open-ended subsidy to a fixed capitated scheme and the abolition of the planning and control system – was to be known as the 'Hiltunen model'. In light of the increasing government budget deficit and the start of the economic recession (which turned out to be the worst since the Second World War), the Hiltunen model was accepted by the government and parliament in 1991, and became effective as of 1 January 1993. The municipalities and their interest groups accepted the plan, as it provided for a considerable increase in their freedom to act and space to manoeuvre.

The Hiltunen model strengthens municipal responsibility for health care by allocating all government subsidies directly to the municipalities. Earlier, subsidies had been paid directly to the service-providing institutions. The municipalities receive a capitated subsidy, and are expected to fund the rest (30–70 per cent of baseline expenditure, depending on economic viability) through local taxes. The subsidy is not earmarked, however, and each municipal council divides its total budget between different service sectors based on its own judgement. It is also free, in principle, to choose the ways in which its 'health care responsibility' is carried out, by either providing services itself or purchasing services, including services in the private sector. In this respect, the new subsidy system makes the municipalities 'purchasers' of health care. The only restriction is that the municipalities still have to remain members of the municipal federation in the province, established to provide hospital services. Membership does not require the municipalities to be purchasers, however.

Providers of specialized care and hospital services lost their government subsidy, and now have to cover their budgets by contracting with purchasers, mainly the municipalities. In principle, providers, like their Swedish counterparts, have shifted from a budget economy to one dependent on performance-related earnings. The public provider units are still run by the municipal federations, although they are relatively flexible in how they organize their internal management due to a recent change in the Local Self-government Act. Prior to the reform of government subsidies, an administrative 'rationalization' was performed by merging all previous hospital-specific federations within a province into one regional organization, thus reducing the number of federations from nearly one hundred to twenty-one. This will be referred to later as the 'administrative reform', which is to be differentiated from the 'subsidy reform'.

Table 4.1   Comparison of health care reforms in Sweden and Finland

	Sweden	*Finland*
Initiation	Locally, within county	From the centre, by Department of Finance
Aim	Higher efficiency and cost containment	To control government subsidies and change incentives for the municipalities
Organizational change	Purchaser–provider split within county	Administrative rationalization
Instruments and ideologies	Competition, entrepreneurship, freedom of choice	MBO, decentralization, abolition of central planning
The role of politicians	Strategy and policies, consumer representation	Not discussed

In summary, the reform of Finnish health care is basically a change in the government's subsidy system, intended to change financial incentives for local authorities. It was initiated by the centre (i.e. the Department of Finance) and increases freedom at the local level by abolishing the planning system. No changes in the organizational or administrative set-up was included, no reference to a purchaser–provider split was made, and competition was mentioned only on the fringes of the debate. The role of politicians was not included in the discussion.

THE CASE STUDIES

Finland

FI-1

The municipal federation in this rural province runs two local hospitals and a mental health organization in addition to a university hospital. It was an early supporter of management by objectives (MBO) and internal decentralization, and was the first to actively prepare its organization for the change in the subsidy system. In anticipation of the 1991 administrative reforms, the university hospital federation took the initiative by voluntarily merging the five municipal hospital federations in the area. At the same time, the organizational structure was changed, radically decentralizing responsibility and decision-making to the departmental level. It was the first MBO–hospital conglomerate in the country. Its corporate structure was to be that of a 'concern' or corporation, with a small management team and support staff at the top, leaving operational management to independent hospitals. The political organization was unaltered, although the mergers reduced the number of politicians involved in decision-making. The federation has a council of municipal delegates, and a board of appointed politicians. The hospital boards, also comprising appointed politicians, were later to be abolished. When reshaping the organization, ideas were expressed about gradually changing the corporate management structure into a purchasing organization, thus anticipating the subsidy reform. The municipalities in the province are, with few exceptions, small, the average population being 3000–4000, too small to be financially viable purchasers of specialized care. Eventually, the senior managers at the corporation level and politicians on the board decided the federation should remain a provider organization, and to increase its grip on the operational management of the hospitals.

The university hospital continued its preparations for the subsidy reform by launching a costing and information systems project. The objective was to replace the old hospital-average daily invoicing system with a per-case-based system with fixed prices on 'treatment packages'.

The first post-subsidy reform budget for the merged federation was negotiated in the autumn of 1992 in the old way: specifying a total budget contribution per municipality, based on historical utilization data, and including an overall 4 per cent expenditure reduction. The municipalities agreed to pay monthly instalments in advance, with a clearing of accounts to be made later during 1993 based on actual utilization and per-case prices.

The pricing project and the preceding MBO exercise were highly publicized in the province. The chances of the municipalities acting as purchasers were more actively discussed than in other parts of the country. In a preceding survey, the municipality's administrators and primary care centre medical directors expressed an active interest in selective contracting for specialized services (Korhonen 1993). Private providers in the area started offering their services, including outpatient consultations and day surgery. In reality, little 'shopping around' happened in the province. The municipal councils and health boards delegated decisions about arranging health care services for their inhabitants to the municipal health centres and their administrators. No active discussion took place about changing policies or strategies, or about replacing providers. For 1994, the municipal administrators required the federation to cut its budget and to reduce its municipal contribution. The per-case invoicing system was installed several months late, and information about utilization patterns has not yet had an impact on municipal behaviour.

In the province, neither the local health boards nor their supporting administrative staff have adopted a purchaser role. Contracts including specifications regarding activities have yet to replace old decisions regarding budget contributions.

FI-2

This province encompasses Helsinki, the capital of Finland. Helsinki runs a comprehensive health care organization, encompassing primary as well as secondary care services. The hospitals in the rest of the province are administered by a municipal federation. A second federation with the same members plus the capital run the university hospital. FI-2 is a geographical area with 18 hospitals within a radius of 50 km, thus creating scope for competition. There are also two private hospitals in the area, and at present the provision of private outpatient services exceeds demand. The university hospital has the capacity to specify its patient care costs by diagnostic and therapeutic procedures, and the costing system could be relatively easily introduced to other hospitals as well. Each hospital has a political board, engaged in decisions about operations. The same type of MBO-directed delegation of decision-making and responsibility as in FI-1 has not, by and large, been introduced. On the other hand, local hospitals in the provincial federation (outside the capital) have in reality preserved most of their independence despite the administrative reforms of 1991. The local boards with their appointed politicians represent the catchment area municipalities, and are strongly committed to the hospital. In many cases, the same politicians serve on municipal councils or health boards and hospital boards. Hospital board membership has traditionally been considered prestigious.

The municipalities in FI-2 are larger and economically stronger than in FI-1. Several big municipalities with a population of 100,000–500,000 inhabitants form strong potential purchasers. Despite this, the municipal councils or health boards have not initiated discussions about selective contracting. The hospitals are increasingly aware of the evolving competition – in light of the ongoing public sector expenditure reduction scheme, necessary due to the economic crisis – and are starting to market their services. Budgets for 1994 were still prepared according to the old procedures.

FI-3

This province is situated in the archipelago between the Finnish mainland and Sweden, and has an autonomous status within the country. The provincial government has been responsible for the specialized care hospital in the province, whereas the municipalities have run primary care and psychiatric care, administered by two separate federations and subsidized by the provincial government. In 1991, the provincial parliament decided to integrate all health care services under the jurisdiction of its government. A planning committee of provincial and municipal administrators proposed a single combined provider organization for all health care. The provider would be organizationally separated from funding and commissioning. The latter tasks were to be handled by the provincial government and its health department, at that time responsible for environmental health, service planning and inspection of health care professionals.

The new arrangement resembled that of a Swedish county model, and was obviously inspired by Swedish reform plans. It included a distinct split between purchaser and provider organizations within the province, with the provincial government continuing as the owner of the provider units. The need to guarantee certain specified services locally was seen as a non-negotiable provincial policy, thus requiring the province to engage in service provision. The discussions, however, also involved the possibility of purchasing services from private sources. Services had earlier been purchased from providers in mainland Finland and Sweden.

As the merger took place in early 1994, no change in behaviour under the new arrangements can be reported yet. This case is of interest because, unlike in FI-1 and FI-2, the role of the politicians was explicitly discussed in the preparatory phase. Two alternative organizational models were discussed, labelled the 'business model' and the 'administration model', respectively. In the first model, the provider organization would not have had any political board. The chief executive would have been appointed directly by the provincial government, and would be responsible for the financial viability of the organization, as well as for delivering services according to the contract. Under that model, the government health department would have been the purchaser with a direct political responsibility, personified in the health minister. Political decisions concerning policy, health needs and purchasing strategy as well as decisions on contracts would have been the task of the minister, the government and the provincial parliament. Under the second model, the provider organization would have been a part of the government administration, separate from the purchasing health department, and the provider organization would have been a board of appointed politicians.

The provincial government chose the 'administration model' as its proposal for the parliament. Also, conservative ministers, some of whom had a business background, rejected the 'business model', because they felt it was the role of politicians to 'keep control' of how health care is delivered.

Sweden

S-1

The first Swedish case is a county which embarked early upon a reform plan based on a purchaser–provider split. The county was involved in community diagnosis planning work in the early 1980s, and started to adjust the funding of its institutions based on catchment area figures, inspired by the RAWP formula of the British NHS. The need to develop activity and cost information handling was realized, and a major investment in the information infrastructure was launched at the end of the 1980s. The county council had been undergoing a series of reorganizations, including the introduction of primary care organizations with the same boundaries as the corresponding municipalities, and the active involvement of local communities by the appointment of health boards with county politicians from the area. In 1991, the county council decided to separate responsibility for planning and commissioning services from provision, and an implementation plan was adopted. According to its 'model', purchasing would be the task of the local health boards, which were to be supported by the county purchasing administration. The support staff consisted of a central office within the county headquarters and three district offices, responsible for servicing the local boards with information support, technical assistance and preparation of purchasing plans. Of the provider units, the hospitals were 'directly managed', with a chief executive directly accountable to the county management. Primary care centres within catchment areas of local boards were to sign capitated agreements with the board. Since 1993, the hospitals have been reimbursed on a per-case basis, but only up to contracted volumes. An 'income ceiling' for hospitals was set corresponding to 1991 utilization figures. Per-case prices were reduced by 10 per cent to boost productivity.

Implementation of the planned reform in S-1 has been complicated by patients' freedom to seek care across administrative boundaries throughout the region (covering four neighbouring counties). Although patients need a referral to elective specialized consultations and care, primary care has not developed a clear-cut gatekeeper function. It has been difficult for purchasers as well as primary care providers to control patient flows. Some of the local areas have lost patients to hospitals outside the county. As money follows patients, the county has had to cover these losses. After very difficult negotiations, the counties in the region agreed in early 1994 to introduce fixed per-case prices for patients treated outside the 'home county'. Diagnostic-related groups (DRGs) will be used as case definitions in all hospitals. Prices are prospectively set, although still hospital-specific. When previous production volumes are exceeded, only 'marginal cost' invoicing is allowed.

The local health boards in S-1 are well established, they have supervised primary care in the area for several years, and have – despite being appointed by the county council – good local credibility. The boards have yet to act as purchasers, although politicians on the board see their role as such. The reasons are a lack of information support, difficulties controlling patient flows resulting from patients' freedom of choice, and – until early 1994 – the lack of a regional agreement on tariffs. Local politicians on health boards have also declined to make decisions regarding priorities and rationing, and have tried to shift that responsibility to the county level.

S-2

The second Swedish case concerns an urban county council running a large hospital network with a highly publicized overcapacity. S-2 began early to develop its internal management and to follow up on internal decentralization by introducing resource management schemes, and involving clinicians in the management of departments and units. The decision to introduce a purchaser–provider split was taken later in S-2 than in S-1 and another frontrunner county, but the reform was implemented faster and much more vigorously. Per-case reimbursement of surgical specialities was introduced in 1992, and in 1993 all hospitals had to raise their income by selling all their production. A productivity target is included in case prices (by reducing them) in order to force hospitals to cut costs. Hospital boards comprising elected politicians were abolished, and hospital chief executives now report directly to the county council executive committee for health affairs. As in S-1, local health boards comprising politicians appointed by the county council were put in place in order to supervise primary care, psychiatric care and geriatric care in the catchment area.

In 1992, the local boards were given the responsibility of purchasing hospital services for their population, in addition to previous tasks. The area chief executive was to refocus his interest from supervising provision to purchasing. The area administrations received capitated funds to cover expenditure for their basic services as well as hospital services. For 1993, 'population contracts' were signed with the primary care organization in the area, formulating the task in broad verbal terms and setting the budget based on historical costs. Contracts with hospitals concentrated on price. Volume was difficult to fix due to patients' freedom to choose a hospital. The fierce competition between hospitals led to real negotiations over price. As an effort to control volume, the primary care organization was supposed to develop 'cooperation agreements' with local or preferred hospitals. These agreements dealt mainly with treatment protocols.

The activities of the area administrations and the local boards have varied. In many cases, the boards are still heavily involved in managing provision, one reason being the introduction of the family doctor scheme in primary care. In a few of the local boards, politicians have adopted a purchaser role, concentrating on local targets, strategies and negotiations with the providers. The key actor in these cases seems to be the area administrator. The most active local politician was a Conservative Party board chairman, forcefully pressing for privatization of services, using the purchasing power of the board. Most of the political debate was focused on the structural imbalances in the county (overcapacity of hospital services) and proposed hospital closures. Apart from these initiatives, professionals acted with great loyalty and took part in implementing the new administrative schemes. The press published mainly positive reports about the reform and no public outcry followed the reorganizations. According to repeated polls, the population continues to be highly satisfied with health care services, a situation found in other counties as well.

Comparison of the cases

The main difference between the Finnish and Swedish public health care systems, in relation to the research issue, is the different role adopted by elected as against appointed politicians and at which organizational level they can be found. In Sweden, the county councils are elected, and the county is the owner of both primary health centres and hospitals. Local boards with provision or purchasing responsibilities are appointed by the county council. In Finland, municipalities governed by elected politicians are responsible for arranging health care services, and they usually manage primary care centres directly. Secondary care institutions are run by municipal federations, with representatives of the owner appointed to the decision-making bodies. Figure 4.1 shows the comparison in the context of the ongoing reforms.

Organizational relations in the reform models of Swedish and Finnish health care

Purchasing as a task for the political boards in Finland is not yet on the reform agenda and, consequently, has not materialized in practice. In one specific case, a restriction on the political influence to ownership, purchasing and control by contracts was debated by the politicians and rejected. An arrangement directly involving politicians in the management of service provision was preferred.

In Sweden, politicians have – at least in principle – welcomed the possibility of concentrating on policy and strategy issues, including the structure of the service system, and have actively discussed a purchaser role. Elected politicians have agreed to the abolition of hospital boards, but overall they still control the county's health care organization. Purchasing has been delegated to locally appointed health boards, which originally had a supervisory role for primary care. The purchasing task is only gradually materializing, and the contracting process is thus far technical in nature and dominated by administrators.

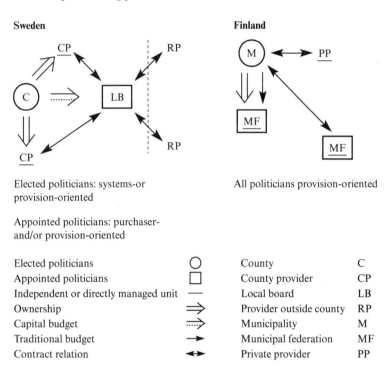

Figure 4.1   A comparison of Finnish and Swedish health care systems

When comparing the two countries, the following differences – apart from the dissimilarities in system and structure reported above – should be borne in mind. In Sweden, the debate was initiated several years earlier than in Finland. Moreover, decentralization on a national scale as well as within health care organizations was started 10–20 years earlier than in Finland. The Swedish reforms, geared at the institutional level, originated inside the health care system and were actively discussed by county politicians responsible for health care. However, in Finland, reform of the system was a genuine change to the funding system, initiated out of concern for the national economy by the Department of Finance. It was assessed by the local politicians based on its direct effect on the size of national subsidies that flow to municipalities, and the increased degree of freedom as a result of dismantling the planning and control system. Politicians involved in health care took part in the discussions as local or national politicians. Health care institutions and their managements were involved in strategic planning for the reforms, but did not actively participate in the design of new 'models.'

The overall impression is that there is still little involvement by the political boards in contracting in Finland, and that the Swedish cases are far less 'advanced' than could be expected in comparison to Finland, considering the system and policy differences.

DISCUSSION

The finding of the overall system and reform programme comparisons, as well as the case presentations, raise a series of intriguing questions. Before turning to them, however, it should be pointed out that all conclusions about the reforms and their consequences, and especially attempts to make causal propositions, are tentative. The reforms are only in their first or second year, and they have not been introduced with the same political decisiveness and rigour as in the nationally controlled 'monolithic' British NHS. Thus all generalizations and concluding statements must be seen as mere hypotheses at this stage.

The lack of far-reaching involvement by politicians and political boards in purchasing is in part dependent on the early stage of the reforms and the lack of an overall political programme, rather than a basic incompatibility of purchasing and political behaviour. However, there is still a case for discussing the purchaser–provider split model in relation to the political decision-making process.

As has been pointed out by Brommels (1991), based on theory, there are several advantages to restricting politicians to a purchaser role only. Brunsson's (1987) distinctions between decision rationality and action rationality, and decision-oriented versus action-oriented organizations, illuminate this argument. Separating service provision from the political decision-making process, and placing it within an action-oriented organization with clear managerial lines of responsibility, is likely to increase the scope for efficient execution of its task. If the task is obscured by a political body representing a number of groups with differing interests, performance is in danger of being hampered. On the other hand, if the political organization is relieved of the responsibility for production, it can be much more flexible in its decision-making and able to involve as many interests as possible in the process. It will then be much more genuinely able to reflect its different constituencies, to increase the representativeness of the public health care organization and, consequently, to increase its legitimacy. This is not a trivial consequence. The widening gap between the growing expectations of the public and the responsiveness of the service organization to customers is eroding the legitimacy of the system. An organizational decoupling would, in theory, both increase the efficiency and responsiveness of the provider organization and strengthen the representative base of the purchaser. In addition, if the politicians on the purchasing body do not have to engage themselves in the trivia of operational management, they will be able to concentrate on their real responsibilities – policies and strategy.

This is the theory. What about the case studies and how do they reflect the reality? The politicians have not actively sought in any of these cases to relieve themselves of tedious tasks connected with service provision, or to withdraw to less controversial and more rewarding strategic decision-making on commissioning boards. This seems to be the case despite the apparent advantages. In Finland, the change in role has not yet been discussed, except for case FI-3, where the politicians actively chose to preserve their direct involvement in provision. In Sweden, purchasing has been delegated to appointed politicians and county staff, whereas the politicians with real power in the county councils have preserved their control over the providing units. Decision-making is at present – as shown in particular by case S-2 – targeted at making direct structural changes in service production. The lesson seems to be that the internal market itself cannot do those 'dirty jobs'. This is a behaviour which, from the theoretical point of view, seems to be contrary to the perceived 'correct' role of politicians.

Let us return to the distinction between elected and appointed politicians. In theory again, one would anticipate that politicians with direct power (i.e. those who are elected) would actively reshape their role in accordance with their interests, and that those who have been appointed would have to put up with tasks delegated to them. If the theory is right, we would find elected politicians creating commissioning bodies for themselves, and leaving the tedious operational management issues to appointed officials. In Finland, the elected *municipal* politicians now have the power – and also the incentive – to adopt a purchaser or strategist role. Reiterating the fact that the reform is in its early phase and that the reform did not emphasize the organizational arrangements as in the Swedish cases, it is notable that there is no active interest in changing the role of politicians. In case FI-1, the idea of reshaping the corporate management of the municipal federation into a purchaser organization was actively rejected. The politicians in FI-3 specifically wanted to preserve their direct influence over the provider organization. In Sweden, the county politicians have been ready to delegate *purchasing* to appointed local boards, but the internal market remains regulated by the county, and they continue to control the structure of service provision. Purchasing does not seem to be the favourite role of the elected politician.

How does one understand this resistance among politicians to adopt the role of the policy-maker/strategist? Is it compatible with the widely accepted goals of the reform, which are to increase the efficiency and responsiveness of health care?

The Finnish and Swedish reforms have one common goal, *decentralization of the public sector*. The market-oriented reforms are superimposed upon a more long-term developmental trend, the shifting of power from the centre to the periphery. One important way of doing this has been to strengthen the role of local authorities, and to reduce central government control. The decentralization process and its consequences can be illustrated by the so-called 'free municipality' experiments, launched in all the Nordic countries in the 1980s. In those experiments, the municipalities were allowed, by special rules and regulations, to organize their activities with fewer legislative norms and less central control. The result in terms of what organizational changes were achieved are amazingly uniform across the participating municipalities, regardless of country: municipal administration was streamlined. In practice, it meant that the number of political bodies – committees appointed earlier by law – was reduced by combining their authority and tasks, thus reducing the number of politicians actively involved in decision-making. At the same time, decision-making power was delegated to administrators, in the name of greater managerial efficiency. The result was a shift in power from politicians to administrators, and a professionalization of municipal services (Ståhlberg *et al.* 1992). However, the politicians are also moving towards becoming *professional politicians* (Stolpe 1992), full-time rather than voluntary workers, and to be increasingly integrated into the administrative machinery.

Rather than to underline the difference in 'success' between the free municipality experiment and the health care reforms, or to try to interpret the politicians' resistance to health care reform as a reaction to losing their positions in the municipalities, the example can be used to highlight the existence of different domains within public organizations (Kouzes and Micro 1979). Administrators and professionals are not 'neutral' or 'passive' actors, only implementing policies and assessing alternative actions for decision-makers. They are actively involved. It is more fruitful to interpret the reforms as an organizational evolution, formed in a complex interplay between a set of actors with different interests, rather than the implementation of a master plan, designed in a grand moment of insight, and then meticulously implemented by loyal administrators.

The actor or stakeholder view and, as a consequence, incrementalistic decision-making, seems to be a plausible model of what is going on in health care, considering its value for citizens and the great interest it raises among the actors involved (Grund 1991). Because of the involvement of a great number of stakeholders, health care would seem – regardless of whether it is organized as a public organization or on a private basis – to be the *archetype* of the political organization, as defined by Brunsson (1989). As most citizens have direct contact with health care, or indirect contact through a family member, friend or neighbour, health care is high on the political agenda. Elected politicians are made accountable, not only when the policy goals of equal access or prioritizing preventive services are met, but even more so for the health centre telephone line being busy or a relative having been treated with arrogance by the hospital staff. Will such politicians give up their direct influence on how services are provided? Not if they have any survival instinct!

From this perspective, elected politicians behave totally rationally when they are unwilling to give up their direct influence on operations and to adopt a mere strategist (purchaser) role. They are much less ready than administrators to discuss the virtues of contract management, and how, by empowering service organization employees, it will accomplish higher customer responsiveness. Their own job is directly at stake.

Why have politicians agreed to be involved in the reforms in the first place? Is it all double-talk? The answer is certainly negative. Politicians undoubtedly accept the need for urgent change in health care. They have actively been promoting new management models, from decentralization and MBO to contract management as measures to enhance efficiency and contain costs. Reforms themselves act as legitimizing strategies; by initiating reforms, the willingness to change is demonstrated to those unhappy with the services (Brunsson and Olsen 1990). The politicians have an objective interest to support the reforms. It would be naive, though, to ignore domain theory and the stakeholder view. Any reforms will activate interest groups, which will block radical change, threatening to influence the existing balance of power. It will take a strong political will – as in British 'conviction politics' – or a painstaking consensus-building effort to achieve real change.

In the Swedish as well as the Finnish setting, the purchaser–provider split will affect the position of elected politicians. In Finland, politicians will have to withdraw from provider organizations, and use their influence solely in municipal commissioning. It is doubtful whether they are willing to do so. In a Swedish county, purchasing can be delegated to appointed politicians and administrators, as long as the elected politicians as owners have ultimate control over the providers.

It needs to be added, though, that control in this instance is used in the formal sense of the word. It is outside the scope of this study to consider the extent to which politicians have been able to influence medical professionals and their pattern of practice. The new incentive schemes introduced as a part of the reforms undoubtedly have had an effect on the way in which provider organizations work. The S-2 case study notes up to 10 per cent increases in productivity during the first year of per-case funding. To preserve such an achievement, provider managers might argue that their autonomous status needs to be retained, meaning that politicians need to be kept out of hospital management. If these 'public firms' (Saltman and von Otter 1992) in an internal market with free patient choice succeed in winning the direct support of citizens and thus securing public legitimacy for themselves, their managers may well develop into powerful actors in the political arena as well.

With the reforms in their first two years, the empirical evidence is not yet available to test the theory. The next opportunity to gain additional insight might occur in Sweden. The position of the county politicians will be strongly challenged if purchasing budgets are shifted to local authorities outside the control of the county as proposed in the 'primary care directed funding model' by a Department of Health task force (SOU 1993: 38). The theory would then predict that county politicians will try to regain operative control of provision.

Of the two basic features of Nordic health care underlined in the Introduction – its comprehensiveness and public accountability – only the latter has been addressed by this study. What are the lessons learnt for other countries? The case studies show that public accountability in the setting of representative democracy keeps elected politicians heavily involved in decision-making, independent of such new administrative arrangements as the purchaser–provider split. Politicians respond to direct pressure from their constituencies. They are not willing to give up control of the public resources they are accountable for. Introducing direct consumer choice does not reduce their sense of overall responsibility. They want to preserve their ability not only to regulate the internal market but also to intervene in decisions at an operational level. Structural changes (e.g. reductions in overcapacity) seem to require the active involvement of the politicians. Politicians involved in running publicly owned health services are in a position to act out of concern for the whole population, not only to respond to those citizens actively using the services.

The theory about decoupling purchasing (aimed at securing legitimacy) from providing (giving it freedom to organize efficiently) got into trouble when looking at the Nordic situation. Politicians feel they can't let go. The theoretical arguments in favour of the reform model have not been refuted, however. The lesson to be learned from this study is not that it will *never* happen, but rather that it will not happen if the rules of the political game are not understood, and unless a sufficient number of actors truly see that their interests – along with those of the population as a whole – will best be served by separating political from professional responsibility.

REFERENCES

Brogren, P.O. and Brommels, M. (1990). 'Central and local control in Nordic healthcare: The public organisation spectrum revised'. *International Journal of Health Planning and Management*, 5, 27–39.

Brommels, M. (1991). 'Spelet om hälsopolitiken'. In M. Brommels (ed.), *Resultat, kvalitet, valfrihet. Nordisk hälsopolitik på 90-talet*. Copenhagen: NOMESKO.

Brunsson, N. (1987). *The Irrational Organization*. Chichester: John Wiley.

Brunsson, N. (1989). *The Organization of Hypocrisy*. Chichester: John Wiley.

Brunsson, N. and Olsen, J.P. (eds) (1990). *Makten att reformera*. Stockholm: Carlssons.

Crossroads Review (1991). *Vägval. Hälso- och sjukvårdens övergripande struktur och framtiden*. Stockholm: Landstingsförbundet.

Gabrieli, B., Leissner, U., Lysell, E. *et al.* (1990). *Decentralisering hela vägen.* Lund: Studentlitteratur.

Grund, J. (1991). *Helsepolitikk i 1990-årene.* Otta: Tano.

Håkansson, S. (1986). 'Frame budgets in Sweden'. In *International Hospital Federation Yearbook*, pp. 189–92. London: Sterling Publications Group.

Korhonen, T. (1993). 'From command-and-control to markets? The future of the Finnish health care system'. Paper presented at the *European Healthcare Management Association Annual Conference*, Warsaw, 29 June–2 July.

Kouzes, J.M. and Micro, P.M. (1979). 'Domain theory: An introduction to organizational behavior in human service organizations'. *Journal of Applied Behavioral Science*, 15, 449–69.

Maxwell, R.J. (1981). *Health and Wealth: An International Study of Health-care Spending.* Toronto: Lexington Books.

Rombach, B. (1991). *Det går inte att styra med mål.* Lund: Studentlitteratur.

Saltman, R.B. and von Otter, C. (1992). *Planned Markets and Public Competition.* Buckingham: Open University Press.

SOU (1993). *Hälso- och sjukvården i framtiden- tre modeller.* Stockholm: Allmänna förlaget.

Ståhlberg, K. (1992). 'Professiokrati och förvaltnung.' In C. Stolpe and K. Ståhlberg (eds), *Professioner, politik och förvaltning.* Åbo: Åbo akademis tryckeri.

Stolpe, C. 'Kan politiken vara professionell?' In C. Stolpe and K. Ståhlberg (eds), *Professioner, politik och förvaltning.* Åbo: Åbo akademis tryckeri.

PART II

BALANCING INCENTIVES
AND ACCOUNTABILITY

5

COSTS, PRODUCTIVITY AND FINANCIAL OUTCOMES OF MANAGED CARE

Nancy M. Kane

Planners, purchasers, providers and patients are the basic players in the health systems of all nations. However, the relative power, responsibilities, sponsorship and goals of each player vary greatly.

In the USA over the last two decades, planning was done primarily by private providers, and served primarily provider goals. Purchasing was represented by a pluralistic system of public and private insurance companies and employers, each with limited power to influence providers. Providers, particularly physicians, were the sovereigns of the system, accountable only to their patients. Patients tended to be compliant, with relatively little basis for exerting independence within their chosen doctor–patient relationships; however, patients were free to exit those relationships and start new ones whenever they wished.

In the USA, as in Sweden and other industrialized Western nations, reform movements are underway to adjust the balance of power among the four players of the health system. Most of the systems have been in place for several decades, and were designed or evolved to meet the needs of a post-Second World War era. Technological change, demographic shifts, rising consumer expectations, erosion of the doctor–patient relationship and rising health care costs have contributed to the climate for reform.

The USA has also experienced changes that seek to reform the role of purchasers of care, and to strengthen purchaser incentives and power to influence providers. The US reforms have been largely private-sector-driven, because its purchasers are largely private sector organizations. The set of reforms underway are collectively referred to as 'managed care'.

While the role of purchaser is under debate in many Western countries, the likely success of any reform mechanism is contingent upon the historical and cultural context in which it is applied. The seeming paradox that strengthening the purchaser role through a physician gatekeeper mechanism is seen as limiting consumer freedoms in the USA, while the very same reform in the UK is seen as increasing consumer leverage over the system, can be explained by the differences in system structure and in historical context. This chapter seeks to clarify the problems as well as the achievements of managed care in the USA to date.

INTRODUCTION TO MANAGED CARE IN THE USA

In the 1980s, 'managed care' became a mainstream health policy alternative to 'regulation' as a means of reforming the health care system and containing health care costs. As perceived in the USA, 'regulation' represents *external government intervention* in the private production of medical services, largely through institutional rate regulation, individual provider fee schedules, authority to deny unnecessary capital expenditures, and limitations upon benefit coverages. Regulatory programmes were widely adopted in the 1970s as the means to control escalating health care costs. However, their political attractiveness was undermined by perceptions, vigorously advanced by the regulated industry, of being rigid, bureaucratic, non-responsive control systems that might curb abuses but also stifled private initiatives to achieve fundamental change.

'Managed care', in the USA, represents *private sector management* of privately organized health networks, in which the goals of cost-effective care are achieved by 'inducements', flexible methods to 'internalize' incentives for cost control in a pluralistic system (Freund 1987; Moran and Wolfe 1991). Ideally, managed care represents 'discretionary and unique decisions to influence care through an interactive process with patients and providers', in order to fundamentally alter the way medicine is provided (Berenson 1991). However, in practice, any attempt on the part of the purchasers of services (employers and insurers) to manage directly the delivery of health services to a defined group of enrollees is called 'managed care'.

Managed care is not managed competition

Managed care has come to include a set of elements, at least one of which must be present to qualify a programme as 'managed'. However, the market structures within which managed care programmes are undertaken may or may not be competitive, and may or may not involve a strong government role. In evaluating the performance of managed care programmes, it is important to distinguish between the effects attributed to the various programme elements, versus the influence of the market structures within which they operate.

The USA has experimented with a wide variety of managed care programmes in an almost equally varied range of market settings. One of the rarest combinations to be found in the USA is competition among clinically integrated networks of providers, or 'the real thing' – that is, managed competition (Freund 1987; Luft 1991; Moran and Wolfe 1991). One journalist recently described managed competition, in the context of President Clinton's proposals for national health reform, as 'a health care theory, largely untested, based on free-market competition. Groups of providers would compete to offer services, and consumers would choose health plans based on which offers the best quality for the lowest price' (Neuffer 1993: 1).

A number of obstacles to realizing true managed competition have been identified. In the private sector, where purchasers are employers, one of the most significant obstacles has been the difficulty employers have in evaluating the basis of 'competitive' prices among traditional 'indemnity' insurance carriers and managed care plans. A large factor in price differences has been attributed to selectivity bias – that is, the tendency of the more traditional managed care plans (health maintenance organizations, or HMOs) to attract the healthiest members of an employee group. This bias is exacerbated by the pricing practice differences between HMOs and traditional indemnity plans; HMOs charged the equivalent of an 'average' price for all employer groups in the geographic community, while the alternative, indemnity plans, charged prices reflecting the actual costs (or experience) of those who enrolled in the indemnity plan. Thus employers with multiple plans often paid more than they should have for the healthier people (whose costs were below the average), and lost the opportunity to pool the healthier people in the same experience pool with the sicker people, thus paying above average costs for their indemnity plans. This problem has motivated an increasing number of employers to select *one insurance carrier*, which in turn offers all employees the 'choice' of its indemnity, preferred provider or health maintainance organization option. This is a distinctly non-competitive trend from the consumer (employee) point of view.

On the consumer side, a number of impediments to true free market competition also remain: the lack of economic motivation for most individuals to choose the least-cost plan, the lack of information with which individuals might assess health care quality, and the unwillingness of people with existing medical conditions or established physician relations to accept restricted provider choice. Finally, there is the problem of scale: it is not clear that rural populations can support more than one efficient network of providers. Also, for certain highly specialized services such as rehabilitation, mental health and acute tertiary referral services, the 'efficient' service area may be national or regional in scope. Thus the notion of competing, exclusive, full-service provider networks seems unlikely to ever be fully realized (Moran and Wolfe 1991).

Ironically, it is probably necessary to increase the role of *government* in order to most fully realize the benefits of managed care. During the most recent federal health policy reform debates in the USA, policy-makers came to believe that there were some functions that were most effectively accomplished by a central rule-maker (the federal government). These included reforming insurance markets to eliminate the opportunities for competition on the basis of favourable selection; developing 'standards' of acceptable medical practice; developing uniform data requirements, centralized collection, and wide dessemination of reports regarding costs and outcomes of delivery system performance; redirection of the training of medical professionals to correct the current imbalance of excess specialists and too few generalists; and reformation of tax policy to make the consumer more conscious of the financial implications of specific health plan choices. The Clinton Administration's proposals went one step further, to define the role of government as price regulator as well. This was one of the most controversial aspects of the Clinton Administration's reform proposal: bringing in government to set overall price caps that could not be exceeded in each geographic area. The focus of this chapter, however, is on current experience of managed care.

THE ELEMENTS OF MANAGED CARE

With or without competition, managed care dominates the private insurance market in the USA: 95 per cent of employment-based insurance is now considered to contain elements of 'managed care'. Managed care is less widely used in public insurance programmes. As of 1991, only thirty-one states had adopted some managed care features for their Medicaid (insurance for the poor) programmes, and only slightly over 2 million of Medicare's 30 million elderly beneficiaries have enrolled in managed care programmes.

The elements of managed care include the following:

1 *Limited choice of provider*. For managed care plans with this feature – the 'health maintenance organization' (HMO) option and the 'preferred provider organization' (PPO) option – the enrollee is free to choose a primary care physician (PCP) within the group of participating providers. The enrollee may be totally responsible financially for any unauthorized care provided outside the participating provider group (HMO option), or the enrollee may be only partially financially responsible (PPO option). Some HMO models, particularly Independent Practice Associations (IPAs) sponsored by medical foundations, soften the extent of limited choice by obtaining participation agreements with nearly every provider (physician and hospital) in a geographic area, but this generally has an adverse effect upon cost containment.

2 *Selective contracting*. This refers to contracts between purchasers (health plans or employers) and providers (institutional and/or individual), in which greater patient volume is promised to specific providers who agree to provide services under contract. The contracts generally require that providers cooperate with certain utilization controls and provide services for a discounted price or fee schedule. For a purchaser or plan with strong local presence (i.e. over 5 per cent of a particular provider's revenues), the fewer providers it contracts with, the better the discounts. However, very selective networks represent more limited choices for enrollees, which can adversely affect the plan's market appeal or employee satisfaction.

3 *Financial incentives for providers*. A number of health policy experts believe that the key to successful 'managed care' is in the way financial incentives are crafted for providers, particularly for primary care physicians. In the USA, the physician is believed to

control 70–80 per cent of all health care expenditure decisions, even though physicians themselves receive only 20 per cent of national health expenditures.

Managed care plans use three generic methods to pay physicians: fee-for-service, salary and capitation (an amount per enrollee). However, the 'art' of managed care lies in the numerous ways in which financial risk-sharing is enhanced or moderated through 'withholds' on fee-for-service or capitation payments, bonuses, stop-loss limits, size of risk-sharing group, proportion of the physician's practice affected by the payment arrangement, the specific services for which the primary physicians are at risk, and benefit design (particularly the financial incentives of the patient). A 1987–88 survey found that between 60 and 70 per cent of HMOs use withholds from the primary care physician's capitation or fee to cover deficits in hospital or specialty referral accounts (Pauly *et al.* 1990). Nearly 80 per cent share surpluses from the referral or hospital accounts with PCPs. Roughly one-third share surpluses with specialists. Fewer than 20 per cent of HMOs (mostly for-profit HMOs) put physicians at risk on an individual rather than a group basis. Roughly one-third offer incentives based on productivity.

4 *Gatekeeping.* Another central tenet to 'managed care' is the concept of primary care physicians as gatekeepers: generalist physicians who must authorize all patient access to specialists and hospital care. In the USA, eligible generalists would include internists, paediatricians, family practitioners and, sometimes, obstetricians/gynaecologists. Part of the gatekeeper responsibility is to provide 24-hour access (generally by 'phone) so that patients do not end up in the emergency room after doctors' office hours. Gatekeeping may include financial incentives such as capitation for specific services within gatekeeper authority; however, some plans simply pay the gatekeeper a 'case management' fee (i.e. $1.50 per enrollee per month) and continue to pay the gatekeeper on a fee-for-service basis for direct patient care services.

5 *Physician profiling.* This concept is used by some plans, particularly those that do selective contracting or gatekeeping without financial risk to the gatekeeper. Profiling requires a fairly sophisticated management information system that can track all expenses (hospital and ambulatory) incurred by a panel of patients of a particular gatekeeper physician. These cost and utilization 'profiles' are used to provide feedback to gatekeepers

and, at times, to terminate the contracts of gatekeepers whose practice patterns suggest excessively costly practices.

6 *Utilization review.* Rather than incentives and feedback, utilization review is direct management intervention in medical practice. There are three types of utilization review: prospective, concurrent and retrospective. *Prospective* review (also known as prior authorization or pre-certification) involves requiring physicians to request permission to admit a patient for an elective hospitalization or expensive outpatient procedure. The insurer or utilization review company follows a set of internal protocols for determining whether or not to approve the physician's request.

Concurrent review is undertaken during a hospital stay, in which a specific length of stay is allowed by the insurer, and any extensions must be specifically approved by the insurer. Generally, appeals to a 'medical director' of the insurer or review company are allowed; but if a request is denied, then the plan will not pay for the requested procedure, admission or additional days. The physician may still pursue the treatment, but only if the patient is willing to pay for it him or herself. If, in hindsight, it turns out that the patient should have had the treatment or extra days, and is harmed by not getting it, the physician (not the insurer or review company) can be sued for malpractice.

Retrospective review is when the insurer or review company reviews claims after the services have been provided; if the claims fail certain internal edits or protocols, payment to the provider may be denied. While in many plans the provider is required to absorb the financial loss for a retrospectively denied claim, in some plans providers are free to pursue payment from the patient.

7 *Organizational culture, or provider self-selection.* In the older, established HMO staff or group models, provider self-selection and/or organizational 'culture' of conservative practice is the primary 'managed care' device influencing physician practice. In these settings, physicians who believe in conservative practice styles and preventive medicine seek to join staff or group model HMOs in order to be able to pursue their chosen style of practice. Alternatively, some HMOs (such as Kaiser) have been known to hire a significant proportion of new graduates of family practice residencies; these physicians often will have trained within that HMO's clinics and have been inculcated with the organizational values of the HMO early in their medical careers.

Most of the studies of HMO performance undertaken as of

1980 looked only at these early HMOs. Their findings – clear evidence of reduced hospitalization rates which resulted in lower overall medical costs – may be due primarily to the culture/physician self-selection attributes of those particular HMOs. The findings cannot be generalized to the newer forms of managed care described above (gatekeeping, financial incentives, utilization review, profiling, selective contracting).

ORGANIZATIONAL FORMS OF MANAGED CARE

Managed care today is characterized as 'complex combinations of economic incentives, bureaucratic structures, and personalities' (Luft 1991). The range of organizational variation in managed care in some form has grown exponentially throughout the 1980s.

The largest number of private sector enrollees are in 'managed indemnity' plans, now offered by close to 70 per cent of all large employers (Taylor *et al.* 1992). These represent the traditional fee-for-service health insurance plans, with managed care features added in the last decade. The primary managed care feature added is some form of utilization review, particularly prospective review of elective hospitalizations. Providers are still paid fee-for-service, but only for approved services.

The next largest number of private sector enrollees, over 50 million as of 1989, are in PPO-type plans, where enrollees are fully insured for services provided within a 'selective network' of providers (Luft and Morrison 1991). However, if enrollees go 'out of network', they must pay a higher co-payment, a deductible, or some other financial penalty. According to a 1985 survey, the most common method of developing 'selective networks' was first to choose a hospital based on its location, and then to recruit the physicians on the admitting staff of that hospital to join the network (Gabel *et al.* 1986). There are a number of variations on the PPO concept which vary the amount of financial exposure the enrollee has for out-of-network use, and the point at which the enrollee must decide whether to choose the HMO, PPO or unrestricted insurance coverage.

Finally, over 40 million enrollees are in HMOs as of 1991, close to 16 per cent of the entire US population. The largest group, 18 million, are in IPA model HMOs,[1] followed by 11 million in group model HMOs,[2] 6.4 million in network model HMOs,[3] and 4.6 million in staff model HMOs[4] (Marion Merrell Dow 1992).

Table 5.1 Plan structure

Elements	Care plans					
				HMO		
	Managed FFS	HIO	PPO	IPA	Group	Staff
Limited choice	0	0	1	2	3	3
Utilization review	3	3	3	2	1	1
Provider financial incentives[a]	0	0	0	3	2	1
Selective contracting						
physicians	0	0	3	3	0	0
hospital	0	0	3	3	2	2
Physician profiling	0	3	1	2	0	0
Gatekeepers	0	3	2	3	3	3
Organizational culture	0	0	0	0	3	3

[a] Financial incentives tend to be strongly associated with the specific plan structure. For instance, most staff model HMOs pay physicians a salary; most group model HMOs pay physicians on either a capitation or salary basis; and IPAs tend to pay on either a fee-for-service or a capitation basis. IPAs are most likely to add withholds, bonuses and referral/hospitalization risks to primary physician payment schemes. PPOs and HIOs usually pay on a fee-for-service basis, as, of course, do managed fee-for-service indemnity plans.

There are other 'experimental' forms of managed care, particularly in the public Medicaid programmes, which are administered by the states. Some Medicaid programmes have developed 'health insuring organizations' (HIOs) that require all Medicaid beneficiaries to enrol in one geographically based HIO. The HIO, in turn, must contract with any provider willing to accept the HIO's payment and utilization review terms.

Table 5.1 provides a rough overview of how the various organizational forms and managed care elements tend to occur. The numbers represent relative reliance on the particular element, going from 0 (no reliance) to 3 (very heavy reliance).

Clearly, the range of managed care structures makes it difficult to generalize about the effectiveness of 'managed care'. It is more instructive to describe costs and outcomes of specific elements of managed care. In some of the research reviewed below, the specific elements of managed care were directly evaluated; in other studies, a specific organizational structure (i.e. HMO, PPO) was evaluated, in which case one might attribute the results to the combination of elements associated with that model, according to Table 1. A review of the recent literature on managed care outcomes by elements is given below.

OUTCOMES OF MANAGED CARE

Provider financial incentives

Historically, nearly every study that has compared unmanaged fee-for-service care with alternative delivery systems has found much higher hospitalization rates in the unmanaged fee-for-service plans; the higher hospitalization rates yielded medical expenditures per enrollee that were roughly 10–20 per cent higher (Wallack 1991). However, further research was necessary to attribute the reductions in hospitalization rates to specific managed care elements within the alternative systems.

A comparison of hospitalization rates after adjusting for patient mix characteristics (socio-demographic characteristics, chronic disease and its severity, co-morbid conditions, and health status characteristics) found that fee-for-service practices in solo and single-specialty groups had hospitalization rates over 40 per cent higher than those of prepaid group practices (generally on salary or capitation) (Greenfield *et al.* 1992).

A recent study of differences in performance *within HMOs*, focusing on variations in physician payment, found that salaried and capitation payment arrangements were associated with lower hospitalization rates than were fee-for-service (generally with a withhold) payment arrangements (Hillman *et al.* 1989). However, a follow-up study noted that the financial incentives of the plan were significantly related to lower hospitalizations *only when the plan was for-profit* (Pauly *et al.* 1990). Non-profit plans were quite similar to for-profit plans in their use of specific financial incentive devices, but the devices did not explain differences in hospital use rates among non-profit plans. The authors concluded that for-profit ownership may 'enhance the power (or the need) of management to offer effective rewards for parsimonious use of health care resources'. However, since the study did not adjust for health status of enrollees and the potential for biased selection, it also seems possible that for-profit status might be more strongly associated

with attracting healthier enrollees, thus achieving lower hospitalization use rates through selection rather than specific financial
incentives.

On the outpatient side, a number of interesting findings emerge
from the same studies. Hillman *et al.* (1989) found that higher
primary care visits per enrollee were associated with a higher
percentage of HMO enrollees among a physician's patients, as well
as with placing primary care physicians at risk for the cost of
outpatient tests. Greenfield *et al.* (1992) found that prepaid group
practices had significantly higher office visits and fewer prescriptions per enrollee than did fee-for-service practices. A third study
found that, within the same practice using the same physicians,
hypertensive patients under capitated payment arrangements received fewer lab tests than did hypertensive patients under fee-for-
service payment arrangements, with no differences in one-year
clinical outcome (Murray *et al.* 1992). All three studies suggest that
under salaried and capitated payment arrangements, physicians will
substitute more frequent primary care for more expensive care,
and/or will reduce unnecessary use of outpatient testing. However,
a fourth study noted that primary care physicians under capitation
who are at risk for primary services only will have a strong incentive
to increase referral to specialists, since they can reduce their level of
effort at no risk to themselves (Stearns *et al.*1992).

Another problem was noted in an evaluation of Medicaid
capitation demonstrations: while capitated physicians did indeed
reduce the use of emergency rooms and visits to specialists, the
savings accrued entirely to the primary care physicians because
their capitation rates were based on a formula related to historical
expenditure levels under fee-for-service (95 per cent of fee-for-
service rates) (Merlis 1993). The issue of who gets the benefit of
reductions in resource use brought about by provider payment
incentives clearly is an important consideration, and how it is
resolved can affect the potency of the payment incentives themselves.

Gatekeeping

Even fewer studies have assessed the 'pure' impact of gatekeeping
upon medical utilization patterns. Gatekeeping without financial
risk-sharing was evaluated in Medicaid experiments undertaken in
the 1980s (Hurley *et al.* 1989). In one experiment evaluated,
gatekeepers were paid a gatekeeping fee per enrollee per month, as

well as fee-for-service for any services they provided to enrollees. The evaluation found that gatekeeping did not reduce inpatient use; however, that could be attributed to the fact that the enrolled Medicaid population – mothers and children on welfare – rarely used the hospital for anything other than childbirth. Gatekeepers did suceed in reducing emergency room use overall and in the number of visits to specialists by their adult enrollees. However, mean expenditures per enrollee did not decrease, because the gatekeeper provided more services directly, and administrative costs were high. The authors conclude that, 'in the absence of financial risk, [gatekeeper] programs are now widely recognized as not having the potential to save money. Rather, these programs are seen as a reform strategy that may enhance access, improve continuity, and at least not contribute to cost increases' (Hurley *et al.* 1989). A number of questions remained regarding whether or not the quality of care in gatekeeper systems was adequate, particularly in situations where enrolment in gatekeeper systems was mandatory (only likely in Medicaid programmes).

Selective contracting

The literature on selective contracting without any form of financial risk-sharing with providers provides no evidence that physicians' practice patterns become more efficient. An evaluation of one PPO that only used selective contracting (without gatekeeping or financial risk-sharing) found that cost per enrollee actually increased because of the expansion of outpatient benefits in the PPO plan relative to the traditional indemnity plan offered to that group. Increases in outpatient utilization more than offset the value of the negotiated discounts achieved under selective contracting (Zwanziger and Auerback 1991).

Limited choice

No studies were available that specifically isolated the impact of limited choice of provider upon use of medical services. However, only two plan structures have traditionally required truly limited provider choice: staff and group model HMOs. Limited choice, in these models, is combined with organizational culture and gate-keeping.

The literature on staff and group model HMOs almost universally supports the lower hospitalization rates achieved by these models, even when selection is taken into account. One study estimated that if everyone in the USA were in a staff or group model HMO, national health expenditures would be 10 per cent lower than they are (Staines 1993).

Physician profiling

This method of controlling physician behaviour is also not used by itself. A 1987 survey found that only 44 per cent of HMOs (mostly IPAs) used physician profiling to either select physicians to contract with, or to educate physicians and weed out poor performance. The proportion of PPOs using profiling was only 22 per cent.

However, physician profiling, coupled with a gatekeeper responsibility, was found to be a powerful control combination in a Medicaid demonstration in Detroit. Gatekeepers were paid a monthly gatekeeping fee, plus fee-for-service for services delivered to their enrollees. However, if gatekeeper profiles exceeded the state's guidelines, the physician was first warned, and then expelled from the programme. That programme achieved substantial reductions in costs, 13 per cent below the state's average Medicaid cost per enrollee, and 21 per cent below the state's average cost for mothers and children on welfare (Merlis 1993). The availability of a sophisticated management information system that tracked physician utilization was essential to the success of the physician profiling activity, and is unfortunately not a common feature of either Medicaid or private sector managed care programmes.

Utilization review

Studies of utilization review *before* the explosion of managed fee-for-service suggested that these programmes did little to alter ambulatory or inpatient costs. However, the sophistication of protocol development and claims information systems has greatly improved over the 1980s. Thus more recent studies attribute a one-time reduction in hospitalization costs to vigorous pre-certification and concurrent review programmes, but no impact on the annual rate of growth in costs.

The most scientific evaluation of utilization review was undertaken on the insurance claims of 222 employed groups of one major private insurer, over the period 1984–86 (Feldstein *et al.* 1988). Eighty-eight groups were subject to mandatory pre-admission certification and concurrent review, while 134 groups had no mandatory utilization review. The groups subject to utilization review had 12 per cent lower hospital admissions, 8 per cent fewer inpatient days, 12 per cent lower hospital expenditures and 8 per cent lower total medical expenditures. Utilization review was found to be much more effective upon groups that were previously high users of medical services: savings of up to 30 per cent of hospital expenses were realized. However, for previously low user groups, utilization review had no measurable impact. Analysis of claims trends over time found no impact of utilization review upon the rate of increase in medical costs per enrollee, once the one-time reduction of utilization review occurred.

The federal utilization review programme for Medicare admissions, the peer review organizations (PROs), conducts some prior authorization activity. An analysis of PROs during the early phase of the Medicare prospective payment system (1984–85) found that the PRO programme was responsible for a reduction in Medicare inpatient admissions and a significant reduction in costs (Ermann 1988).

The Congressional Budget Office estimated that if everyone in the USA who was not in a staff or group model HMO was subject to mandatory prior authorization and concurrent review of inpatient care, national health expenditures would fall by 1 per cent (Staines 1993).

SUMMARY OF THE EFFECTIVENESS OF MANAGED CARE ELEMENTS

The most effective managed care combinations are those that include limited choice, organizational culture and gatekeeping. However, Americans do not accept limited choice, as is evident by the 1991 enrolment levels: only 4.6 million have enrolled in staff model HMOs, and only 11 million in group models (excluding network HMOs, which seek to offset limited choice by enrolling not only groups but individual physicians). This represents less than half of all HMO enrolment, and only 16 per cent of the population has enrolled in an HMO of any model. The elderly population (Medicare), in particular, reject limited choice. Less than 5 per cent of the Medicare population has enrolled in an HMO, and those that do tend to be of lower income and not yet in an established relationship with a physician. Thus, while limiting choice to conservative-type practices appears to be associated with more cost-effective use of hospitals, it is not a politically feasible option in the USA at this point. If financial incentives to enrollees are greatly changed, such that beneficiaries had to pay out of their pocket, in after-tax dollars, for all of the difference between a limited choice and an unlimited plan, more people might choose limited choice. However, it is not clear that limited choice alone would achieve all of the benefits of group and staff model HMOs; in fact, provider selection/organizational culture are likely to play a significant role as well. In the USA, there just may not be enough conservative practice-minded physicians to take care of everyone.

The most popular (in terms of enrolment) form of managed care is utilization review. While this allows unlimited provider choice, it puts the provider in the difficult position of having to ask an external agency for permission to admit patients and to keep them in the hospital for additional days. Utilization review activities are a major source of animosity and added administrative expense for providers and insurers. Rather than approaching the ideals of managed care (internalized incentives, interactive process with patients and providers to influence care), utilization review is viewed as a highly visible, regulatory type of intervention in medical practice. Physician surveys indicate that those not in HMO salaried or group practices (where utilization review-type protocols have largely become internalized to practices) find utilization review to be a major source of dissatisfaction in their work life (Baker and Cantor 1993).

Other elements of managed care either don't work in the current environment or present ethical dilemmas, including selective contracting and 'creative' financial incentives for providers. Selective contracting is a difficult notion to operationalize, given that most private insurers represent less than 1 per cent of any specific provider's patients. With that degree of fragmentation, it is difficult to imagine how sufficient volume can be directed to providers to be influential in both price determination and 'interaction' over practice patterns. Private physicians in some markets have contracts with 100 different plans. For selective contracting to be more effective, fewer and larger contracting plans are necessary in a particular market area. This could be accomplished under health reform if politicians can agree to limit the number of plans offered by purchasing groups (employers or alliances) to employees; but the insurance industry has already mounted serious opposition to meaningful limitations.

Financial incentives for providers, especially capitation, suggest some rewards but some risks; the evidence on capitation to date leaves us wondering how those with the greatest capitation risks (primary care physicians at risk for referral and hospitalization as well as their own primary care services) learn to 'manage' that risk. The easiest way to manage it is to dissuade your sickest patients from joining your capitation panel, and limited evidence suggests that this does happen (Hillman *et al.* 1989). On the other hand, capitation that is limited to primary care services only has been shown to encourage increased referrals to specialists, thereby decreasing the primary care physician's workload (Stearns *et al.* 1992).

INFLUENCE OF MANAGED CARE UPON COSTS, PRODUCTIVITY AND OUTCOMES: POLICY LESSONS

Managed care has come to represent the preferred method of allocating health care resources for the private sector in the USA. It also has great appeal to policy-makers concerned with the public programmes of Medicare and Medicaid, although it represents greater implementation challenges in the public sector. What does managed care accomplish?

First, it is apparent that managed care *can* redistribute resources in desirable directions, relative to 'unmanaged' fee-for-service, total freedom-of-choice plans. All forms of managed care appear able to decrease unnecessary hospitalization, and substitute less costly outpatient care when appropriate. In some forms (particularly when the ambulatory benefits are enhanced, as in HMOs and PPOs, and gatekeepers are mandatory), managed care is associated with increased use of primary care physicians, and less use of inappropriate emergency and specialist care.

In addition, numerous analyses of the quality of care suggest that, at a minimum, managed care *medical* quality is at least as good as 'unmanaged care' quality (Ware *et al.* 1986; Greenwald 1987; Luft 1987; Retchin *et al.* 1992; Burns *et al.* 1993). The satisfaction of patients in managed care is often reported to be lower than that of patients in fee-for-service care in terms of *perceptions* of medical quality, as well as accessibility and continuity. However, in instances in which the patients' insurance alternative is a fee-for-service indemnity plan, those in managed care plans are more satisfied with their out-of-pocket expenses, broader preventive coverage, and better coordination of care (Clement *et al*. 1992; US General Accounting Office 1993).

Equally important to policy-makers is what managed care has not yet been able to do. First, it has not been able to lower hospital expenditures per capita, even in markets where the most effective managed care organizations – group and staff model HMOs – are effectively lowering hospital admission rates. According to one national analysis, the only market areas that showed significantly lower growth in hospital expenses per capita were those with mandatory rate-setting programmes (McLaughlin 1988). According to that study, the primary forces driving up hospital expenses per capita were physician supply (more doctors = higher hospital expenses per capita) and the mix of specialists to generalists (more specialists = higher hospital expenses per capita).

Nor has the ubiquitous presence of managed care in the 1980s succeeded in increasing productivity, at least in the hospital sector. Between 1983 and 1987, labour hours per case-mix-adjusted discharge actually *increased* (Altman *et al*. 1990). Meanwhile, reductions in inpatient care have been more than offset by an explosion in the use of outpatient resources.

Finally, while research has firmly established that managed care in its highest form (a well-established staff model HMO) can indeed lower total expenditures per enrollee by 10–20 per cent without lowering health outcomes (Manning *et al*. 1984), it is equally well established that the more commonly available managed care organization is highly susceptible to favourable selection. For the public programmes, particularly Medicare, favourable selection is a major obstacle to voluntary enrolment in managed care, and greatly complicates their efforts to achieve an equitable payment level. A recently completed 4½ year evaluation of Medicare risk-sharing programmes with HMOs found that favourable selection resulted in the Medicare programme paying more for enrollees in HMO plans than it would have paid for them under fee-for-service plans. While the plans reduced lengths of hospital stays significantly, the savings to Medicare were not captured due to the lack of premium adjustments to reflect health status (Brown *et al* 1993). Until premiums can be reasonably well adjusted for health status, favourable selection will continue to undermine the value of managed care for employers and government (US General Accounting Office 1993).

Thus managed care has not slowed the rate of growth in medical expenses in the USA. It has succeeded in redistributing those resources, and in setting the stage for what might be the next step: implementing national health reforms. For managed care to be more effective, the setting in which it operates must change. In particular, purchasers and consumers must become more concerned and more actively involved in securing care at the least cost. Treatment protocols, particularly protocols governing the use of expensive technology, must be promulgated nationally, disseminated widely, and adhered to. Competition over selection of patients, rather than over the efficiency of the delivery system, must be eliminated. In the long run, the mix of generalists to specialists must change. These are all systematic issues that can only be addressed by government action.

To the extent that some form of managed care has appeal in other countries, it must be realized that the various elements alone will probably not achieve the desired outcomes of reduced cost and greater efficiency for the same or better health outcomes. However, if the setting can be properly engineered by government, managed care does offer a vehicle by which risk and responsibility for the care of an enrolled population can be transferred to a number of competing private sector networks of providers. The most powerful managed care elements are limited choice, gatekeepers and an organizational culture that fosters conservative practice. External utilization review – the more bureaucratic/regulatory element – can also be effective, but at the expense of provider autonomy and willingness of physicians to become a part of the solution.

CONCLUSIONS

Clearly, caution must be used when considering the transnational adoption of any of the elements of managed care as they manifest themselves in the USA today. While all of the elements intend to increase purchaser power *vis-à-vis* providers, some elements fit better than others to a particular country context. The primary gatekeeper element apparently fits well in the UK, with its abundance of entrepreneurial general practitioners (Smee 1995). However, selective contracting has been difficult to introduce in the Netherlands, which experimented with deregulation by government and a shift of power to administratively oriented insurance companies lacking the internal expertise to negotiate cost-effective contracts with providers (Van der Kooij 1993). Managed care offers a collection of tools, some of which may increase the power of purchasers to negotiate effectively on behalf of patients, *in the right context*. However, implementation of a tool or mechanism which does not fit the local context can bring health system chaos and political embarrassment to its supporters.

Even in the USA, many voice doubts that managed care can accomplish the goals set out for it in recently proposed national reforms. As proposed by the Clinton Administration in the autumn of 1993, reforms to realize the promise of managed care required significant increases in government planning and regulation. A major increase in government 'intervention' in the health care system is not a politically popular notion in this nation of 'live free or die' individualists. Elements of managed care may have a better chance of success in countries where 'solidarity' permits public sector planning.

NOTES

1 Independent practice associations contract with individual physicians (some of whom may practise in small groups) to provide primary and referral services to enrollees.
2 Group models contract with one or more large multispeciality group practices.
3 Network HMOs contract with individual physicians and group practices.
4 Staff models employ physicians directly on a salaried basis.

REFERENCES

Altman, S.H., Goldberger, S. and Crane, S.C. (1990). 'The need for a national focus on health care productivity'. *Health Affairs*, Spring, pp. 105–13.

Baker, L.C. and Cantor, J.C. (1993). 'Physician satisfaction under managed care'. *Health Affairs*, 12, 258–70 (suppl.).

Berenson, H.A. (1991). 'A physician's view of managed care'. *Health Affairs*, Winter, pp. 106–19.

Brown, R.S., Bergeron, J., Clement, D.G. *et al.* (1993). *The Medicare Risk Program for HMOs: Final Summary Report on Findings from the Evaluation*. Princeton, NJ: Mathematica Policy Research, Inc.

Burns, L.R., Wholey, D.R. and Abeln, M.O. (1993). 'Hospital utilization and mortality levels for patients in the Arizona health care cost containment system'. *Inquiry*, 30, 142–56.

Clement, D.G., Retchin, S.M., Stegall, M.H. and Brown, R.S. (1992).

Evaluation of Access and Satisfaction with Care in the TEFRA Program. Princeton, NJ: Mathematica Policy Research, Inc.

Ermann, D. (1988). 'Hospital utilization review: Part experience, future directions'. *Journal of Health Politics, Policy and Law*, 13, 683–704.

Feldstein, P.J., Wickizer, T.M. and Wheeler, J.R.C. (1988). 'The effects of utilization review programs on health care use and expenditures'. *New England Journal of Medicine*, 318(20), 1310–14.

Freund, D.A. (1987). 'Competitive health plans and alternative payment arrangements for physicians in the U.S.: Public sector examples'. *Health Policy*, 7, 163–73.

Gabel, J., Ermann, D., Rice, T. and de Lissovoy, G. (1986). 'The emergence and future of PPOs'. *Journal of Health Politics, Policy and Law*, 11, 305–22.

Greenfield, S., Nelson, E.C., Zubkoff, M., Manning, W., Rogers, W., Kravitz, R.L., Keller, A., Tarlov, A.R. and Ware, J.E. (1992). 'Variations in resource utilization among medical specialities and systems of care'. *Journal of the American Medical Association*, 267, 1624–30.

Greenwald, H.P. (1987). 'HMO membership, copayment, and initiation of care for cancer: A study of working adults'. *American Journal of Public Health*, 77, 461–6.

Hillman, A.L., Pauly, M.V. and Kerstein, J.J. (1989). 'How do financial incentives affect physicians' clinical decisions and the financial performance of health maintenance organizations?' *New England Journal of Medicine*, 13, 86–92.

Hurley, R.E., Paul, J.E. and Freund, D.A. (1989). 'Going into gatekeeping: An empirical assessment'. *Quality Review Bulletin*, October, pp. 306–14.

Luft, H.S. (1987). *Health Maintenance Organizations: Dimensions of Performance*, Ch. 8. New Brunswick, NJ: Transactions, Inc.

Luft, H.S. (1991). 'Translating the US HMO experience to other health systems. *Health Affairs*, Fall, pp. 172–86.

Luft, H.A. and Morrison, E.M. (1991). 'Alternative delivery systems'. In E. Ginzberg (ed.), *Health Services Research: Key to Health Policy*. A Report from the Foundation of Health Services Research. Cambridge, MA: Harvard University Press.

Manning, W.G., Leibowitz, A., Goldberg, G. *et al.* (1984). 'A controlled trial of the effect of a prepaid group practice on use of services'. *New England Journal of Medicine*, 310, 1505–10.

Marion Merrell Dow (1992). *Managed Care Digest*. HMO Edition. 9300 Ward Parkway, Kansas City, MO 64114, USA.

McLaughlin, C.G. (1988). 'The effect of HMOs on overall hospital expenses: Is anything left after correcting for simultaneity and selectivity?' *Health Services Research*, 23, 421–42.

Merlis, M. (1993). 'Appendix G: Managed care'. In *The Medicaid Source Book: Background Data and Analysis (a 1993 Update)*. Report prepared by the Congressional Research Service for the use of the Subcommittee of Health and the Environment of the Committee on Energy and Commerce, US House of Representatives, January.

Moran, D.W. and Wolfe, P.R. (1991). 'Can managed care control costs?' *Health Affairs*, Winter, pp. 120–8.

Murray, J.P., Greenfield, S., Kaplan, S.H. and Yano, E.M. (1992). 'Ambulatory testing for capitation and fee-for-service patients in the same practice setting'. *Medical Care*, 30, 252–61.

Neuffer, E. (1993). 'Clinton care plan: A policy primer in health care'. *Boston Globe*, 26 August, p. 1.

Pauly, M.V., Hillman, A.L. and Kerstein, J. (1990). 'Managing physician incentives in managed care: The role of for-profit ownership'. *Medical Care*, 28, 1013–24.

Retchin, S.M., Brown, R., Cohen, R. *et al.* (1992). *The Quality of Care in TEFRA HMOs/CMPs*. Technical Report. Princeton, NJ: Mathematica Policy Research, Inc.

Smee, C.H. (1995). 'Self-governing Trusts and budget holding GPs: The British experience'. In R. Saltman and C. von Otter (eds), *Implementing Planned Markets in Health Care*. Buckingham: Open University Press.

Staines, V.S. (1993). 'Potential impact of managed care on national health spending'. *Health Affairs*, 12, 248–57 (suppl.).

Stearns, S.C., Wolfe, B.L. and Kindig, D.A. (1992). 'Physician responses to fee for service and capitation payment'. *Inquiry*, 29, 416–25.

Taylor, H., Leitman, R. and Blendon, R.J. (1992). 'Large employers and managed care'. In R.J. Blendon and T.S. Hyams (eds), *Reforming the System: Containing Health Care Costs in an Era of Universal Coverage*. Washington, DC: Faulkner and Gray.

US General Accounting Office (1993). *Managed Health Care: Effect on Employers' Costs Difficult to Measure*. GAO/HRD-94-3. Report to the Chairman, Subcommittee on Health, Committee on Ways and Means, House of Representatives.

van der Kooij, S. (1993). *Experiences from the Netherlands with an Insurance Operated Healthcare System*. Report prepared for the Swedish Parliamentary Committee on Funding and Organisation of Health Services and Medical Care, HSU 2000, September.

Wallack, S.S. (1991). 'Managed care: Practice, pitfalls, and potential'. *Health Care Financing Review*, 27–35 (suppl.).

Ware, J.E., Brook, R.H., Rogers, W.H. *et al.* (1986). 'Comparison of health outcomes at a health maintenance organization with those of fee-for-service care'. *Lancet*, May, pp. 1017–22.

Zwanziger, J. and Auerback, R.R. (1991). 'Evaluating PPO performance using prior expenditure data'. *Medical Care*, 29, 142–51.

6

VOUCHERS IN PLANNED MARKETS[1]

Richard B. Saltman and Casten von Otter

INTRODUCTION

The issue of risk as it relates to human services is intimately tied to the history of the Western industrial welfare state. The Beveridgean conception of a welfare state was one which collectivized risk, such that separate individuals no longer carried alone the burden for financing potential and/or predictable costs of unemployment, illness or old age. The central objective of the welfare state, at the macro or system level, was to eliminate the individual's risk of finding him or herself with inadequate financial resources to cover necessary costs.

In health care, the role of the traditional welfare state can be conceptualized as one of *risk absorption* and *risk redirection*. In terms of risk absorption, it was intended to reduce financial risks to the individual patient, in particular to vulnerable populations such as the young, the elderly, the chronically ill, and the mentally or physically handicapped. In terms of risk redirection, the role of the welfare state was to spread financial risks for health and social services among the entire citizenry, thereby both reducing the burden on any individual (social insurance) and guaranteeing equal coverage to all inhabitants regardless of class or income (equity, solidarity and social justice).

One could argue that the current interest of some governments in vouchers can be characterized, at the micro or individual service and/or individual citizen level, in quite another manner, as *risk-shifting* and *risk-evasion*. No longer willing or able to continue to absorb the financial risks that it had previously accepted, these welfare states now seek to *reallocate* segments of risk variously to service providers (both insurers and delivery institutions) and to patients. As justification, these governments tend to cite four objectives – micro-economic efficiency, budgetary predictability, individual choice and ideological change – as well as more deep-seated pressures, in particular to reduce overall public sector expenditure and social consumption in an effort to enhance national economic competitiveness. There are also new pressures inside the health system to obtain better value for money and to adapt to individual needs and wants.

The shift in role and strategy – from risk absorption to risk evasion, from socially driven risk redirection to individually oriented risk-shifting – raises a variety of questions about the likely outcomes that could be expected to accompany the introduction of vouchers or voucher-style arrangements in the health care sector. Whether this farming out of micro-level risk is a means to save the welfare state, or a device to speed its dismantlement, depends on how one interprets the balance of these outcomes. In line with the old Roman adage, *cui bono*, one needs to inquire as to who benefits and who loses, and for what reasons.

Over the past several years, policy-makers have sought to apply a range of different competitive instruments to the production side, the finance side, or the allocation mechanism (that connects production to finance) of health systems (Saltman, 1994). One seemingly attractive instrument has been vouchers. Vouchers or voucher-like arrangements have been proposed on the finance side for health insurance by governments in the Netherlands, the USA (for uninsured individuals) and New Zealand. Similar insurance devices have been suggested in Russia and Israel. These proposals all involve establishing a two-tier financing structure, in which a universal insurance pool (tier 1) would be distributed among a variety of competing private insurers (tier 2) on an individual-tied voucher basis. Vouchers on the finance side have been adopted for publicly funded insurance in the USA (for a number of elderly Medicare and poor Medicaid patients), and as an allocation mechanism in several European countries for primary care organized on a capitation basis with independent general practitioners (Britain, the Netherlands and Sweden). On the delivery side of health systems, there have been experiments with voucher arrangements in some municipalities for home care services (Sweden), for other health-related services (including transportation of the elderly), and in some Swedish counties for dental services (von Otter and Tengvald 1992).

To date, vouchers and/or voucher-style arrangements typically have been adopted on an *ad hoc* basis. There is, furthermore, considerable discussion in a number of countries about additional uses for vouchers, particularly for specific bundles of health services. Several academics have suggested that vouchers or voucher-style arrangements can provide an effective solution for the various ailments that afflict publicly operated health systems (Le Grand 1990). However, the implications of such payment configurations for the long-term social as well as financial character-istics of recently unbundled public health systems have not as yet received systematic or theoretically sustained attention.

The role of vouchers as a risk-shifting device needs particularly careful assessment. Presuming policy-makers wish to maintain equity of access to services, certain prerequisites need to be accommodated in the design of a voucher arrangement. The risk pool (relevant population) should be designed so that every subset represents the same profile of high- and low-cost cases. This requires each subset to have a normal distribution, one in which no specific sub-populations will risk creating a bias in the random sampling of subsets. These conditions are rarely fulfilled, as relevant problems (i.e. poor health) often overlap with socially discriminating attributes (gender, ethnicity, income, residential area) to create logistical barriers to 'fair' allocation procedures. To mix patients into similar socio-economic subsets for all general practitioners, for example, would mean assigning people to prac-tices at random across an entire city.

This chapter explores the potential role that vouchers might have in the future development of planned markets in health care. Among the issues considered will be the purposes for which vouchers are used (or suggested for use), past experience with different types of vouchers in different social policy arenas, and various distinctions and constraints that might be applied to vouchers in pursuit of an optimal outcome. The underlying theme of this analysis is the relationship of vouchers to the development of increased accountability of health systems to patients, citizens and professionals. While the focus will be on the use of vouchers within health and related social services (elderly nursing home and home care), a number of the observations could be applied in other human service sectors of the welfare state as well.

CONCEPTUALIZING VOUCHERS

There are two different kinds of vouchers. One is based on a general entitlement with the same nominal sum, intended to equalize purchasing power for a specific service (dental care, maternity care vouchers). This form has recently been introduced in certain counties in Sweden. The other main type of voucher is differentiated according to the recipient's income and/or need for care, calculated either individually or by category.

A voucher is often thought of in literal terms, as a physical piece of script that is distributed to individuals who then can exchange it for a needed good or service. In practice, it is much more likely that a voucher is an accounting category, which defines the amount of money that is transferred to a provider for a particular service chosen (within a certain range) by the entitled individual. Moreover, while patient choice of provider is usually the steering mechanism for this payment, in principle vouchers can be allocated directly by the issuer. In practice, vouchers have been issued almost exclusively by public sector governmental agencies.

Viewed analytically, a voucher can be defined as a prospectively fixed price tied to the provision of a bundle of goods or services, with choice of provider left to the recipient. In essence, it is a type of individualized contract between a funder and a provider. While vouchers are commonly thought of as having a single price for all recipients, in fact they can have multiple prices for different categories of individuals: capitation rates for primary health care in Stockholm County up until January 1994 had two categories, with the rate for citizens over age sixty-five set at double that for those under sixty-five. Similarly, voucher-style arrangements for the two-tiered health insurance model described above can, in principle, be risk-adjusted in order to reduce the incidence of adverse selection by private insurance companies, although recent Dutch experience suggests that it is difficult to develop a risk-rating formula that is sufficiently sensitive to preserve equal access for the entire population (van de Ven 1993).

Vouchers typically are put forward as a means to achieve one or more of four different governmental objectives:

1 *Micro-economic efficiency*: by creating competitive incentives to improve quality and/or lower price among insurance carriers (on the finance side) or providers of health care services (on the production side).

2 *Budgetary predictability*: by fixing the total annual public sector cost for providing either insurance cover or health services in an easily calculable, prospectively budgeted manner.

3 *Individual choice*: by enabling patients or clients to select their insurance carrier or service provider, a process sometimes referred to by voucher proponents as patient empowerment. Working in tandem with competitive incentives, choice is often viewed as a second mechanism to force producers to enhance quality.

4 *Ideological change*: by restructuring publicly provided services to reflect a belief that there are no collective properties to welfare state goods, that only the individual can pursue his or her own subjective interest, and, often by extension, that most or all services should be provided in the private, for-profit sector.

These different objectives tend to be complementary rather than contradictory, with one or another objective highlighted in public debate depending upon the audience. For example, proponents of vouchers typically emphasize the issue of individual choice and micro-economic efficiency in public discussion, while the ability to attain budget predictability and ideological change – often the more central objectives – may be noted only in policy formulating circles.

A key but unstated objective in designing and implementing some human service voucher programmes is to shift financial risk for provision from the public sector onto some combination of the provider and/or the service recipient. Risk-shifting is an inherent aspect of making human service budgets predictable when the need for those services itself is not easily predictable. Risk-shifting can also refer to quality, making the individual at least partially responsible for choosing a provider who will satisfy his or her needs in qualitative as well as quantitative terms. Risk-shifting often lies at the core of how vouchers behave in the health sector, and thus should be treated as an important characteristic of a voucher-based arrangement.

VOUCHERS IN THE WELFARE STATE

While vouchers, until recent years, have had limited use in the health sector, they have a more extensive history in several other welfare state sectors. In the USA, voucher or voucher-type arrangements have been used since 1973 for a supplemental federal

food programme, since 1975 for a federal public housing programme, and since the late 1980s in a growing number of local and
now regional (state) education programmes. By examining the
differing characteristics of the goods and services involved in these
various programmes, it becomes possible to make a clearer assessment of the conditions under which vouchers can be applied, and
the advantages and disadvantages that accompany them.

Voucher programmes for specific goods, for example food and
housing, have four major characteristics. First, individuals supported by the programme have a *fixed need* for the good, which can
be *easily quantified*, readily priced and readily budgeted. An adult
living alone needs so many square metres of space, and so much
food for a day; a family of four requires so many additional square
metres, so much additional food, etc. While one can quibble over
generous or spartan standards, the financial calculation, once
agreed upon, remains essentially constant (if inflation-adjusted).

Second, the goods in question are *physical* in nature. Their
quality can be easily inspected, monitored and evaluated. Being a
physical good, outcomes can be immediately assessed. Third, the
delivery of the good is *one-off*, and does not need to be managed on
a regular daily basis. Fourth, reflecting the predictability of both
need and projected expenditure, *no significant risks are shifted* to
either provider or client by a prospective fixed-price voucher for
these goods. That is, the value of the voucher recompenses both
provider and client for exactly the agreed-upon good (i.e. a certain
amount of food, a certain size of flat).

The provision of human services, by contrast, exhibits substantially different characteristics. First, education, elderly care and
health care all require a *varying quantity* of service, contingent upon
the needs and condition of the individual, which for the latter two
can change rapidly and with no warning. This can make it harder
(education) or extremely hard (elderly care, health care) to predict
accurately the annual need per individual, and then to price and
budget appropriate funds to meet that individual's specific need.
Second, the services in question are *interactive* in nature, defined in
considerable part by the relationship between a patient or client and
a provider (Habermas 1984–7). As a consequence, their *quality* is
intangible and hard to assess. To the extent that it is dependent not
just on formal qualifications but on commitment and enthusiasm,
the quality of these services is hard to inspect, monitor and evaluate. Important outcomes can take months, sometimes years (attitudes towards learning, health status) to observe. Third, adequate

provision entails *considerable management* of service delivery, often on a daily basis (because of the variability of an individual's condition). Fourth, reflecting the unpredictable and intangible character of the good, considerable financial and quality *risk may be transferred* to both service provider and service recipient. If the funds allocated prove inadequate, the provider or the patient must cover the cost of additional services or the patient must go without.

These comparisons suggest that there are important structural differences between establishing a programme of vouchers for specified goods and services (food, housing, education) as against establishing one for interactive and individualized human services like elderly or health care. The demonstrated viability of voucher programmes for specified goods may not be indicative about the likely outcome of an attempt to utilize vouchers for human services that are more difficult to quantify, and that must be managed on an ongoing basis. Furthermore, in terms of programme outcomes, vouchers for specified goods do not shift risk and budget unpredictability away from the public sector onto some mix of provider and/or service recipient such as could occur in the area of elderly or health care services.

VOUCHERS IN HEALTH CARE

Vouchers have been proposed or adopted for use on both the finance and the production sides of health care systems. On the finance side, the voucher typically ties the individual subscriber to a particular insurer for a one-year period, and provides coverage for most clinical medical services. On the production side, vouchers for capitation payments to primary care providers tie a patient to a particular provider for three months (Germany) or for however long the patient wishes (Stockholm County).

When one considers the likely impact of vouchers or voucher-like arrangements on the wide range of services and providers that make up a health care system, that impact would appear to vary in accordance with at least four different types of characteristics: those of the specific clinical service, those of the service provider, those of the bundle of services and those of the patient. An examination of these factors suggests a series of criteria that can help define where vouchers may be more, as opposed to less, useful in designing planned markets in health care:

- *Low or high intensiveness.* For example, primary care as against acute hospital surgery. Generally, services of low intensity present less physical risk to the patient if poorly performed or of low quality than do services of higher intensity.
- *One-time or multiple iterations.* Some services require only a single episode (coronary bypass), while others involve many return visits (elderly home care). The value of being in control of non-medically essential parameters increase with multiple episodes (i.e. dermatology treatments for psoriasis).
- *Easier or harder evaluation of service quality.* For some services, quality can be readily evaluated (corneal lens transplants), while for others it is difficult to determine service quality (intensive home nursing). In some cases, the social chemistry is an important criteria; in others, it has only minimal impact.
- *Sufficient or insufficient information available regarding service quality and outcomes.* For some services, the recipient may have sufficient information to judge the quality of the service or the likely outcomes, while for other services the needed information may be limited and inadequate, or withheld as privileged or proprietary in character.
- *Limited/discrete or all-encompassing service packages.* Some service packages can be simple and transparent, such as many routinized diagnostic tests; other packages are comprehensive and contingent, as for most curative health services.
- *Integrated or fully detached service vendors.* Service providers may be directly owned by the same public authorities, as in some publicly operated health systems, or they may be financially detached, private organizations or individuals. Among the considerations here are whether a provider has 'deep pockets' or not (referring to the provider's ability to offer services below cost over a prolonged period so as to drive competitors out of business and to increase market share), and whether a provider can be sold to another owner, may be subject to a hostile corporate takeover, or can go bankrupt.
- *Services for competent adults or services for socially or mentally incapacitated individuals.* Some services are intended for adults of sound mind, who are competent to judge the nature and quality of the care they need and receive. Others are delivered to patients rendered incapable of choosing effectively due to illness or who are mentally infirm, and who themselves cannot judge an appropriate standard of care. This can be compensated for by

family or trusted guardians, but they may also be easy prey to guileful providers.
- *Public/collective or private/individual benefit.* Most health care services provide benefits to the broader society as well as to individual recipients (Saltman 1993). Examples include not only infectious disease control (i.e. immunizations, AIDS education), but also negotiated service packages that guarantee adequate care to all eligible citizens (i.e. for elderly home care services), the adequate functioning of which is not indifferent to the functioning of society in general, and thus have a collective good aspect.

VOUCHERS AS INSTRUMENTS TO ALLOCATE RISK

Potential benefits

The expected benefits following the introduction of vouchers tend to reflect the official objectives noted earlier. For governments, the use of vouchers can generate fixed expenditures and greater budget predictability and control, especially in the short term. Vouchers can also generate improvements in micro-economic efficiency among service providers. Furthermore, vouchers can serve as instruments of ideological change, helping conservative governments to eliminate the collective character of welfare state services and paving the path toward private provision.

For providers – be they insurers or health care institutions and personnel – one can point to two potential benefits. First, the adoption of vouchers can be viewed as a source of stable income (this only applies to a fee-for-service system like that in the USA, or to private providers outside publicly budgeted structures, as in most Northern European health systems). Additionally, vouchers may be perceived as beneficial by at least some providers in that they herald entrepreneurial freedom – the opportunity to build a bigger practice or business, with a greater degree of day-to-day managerial autonomy.

Lastly, regarding potential benefits for patients, vouchers are expected to provide choice of provider. This is a particularly interesting argument in a society and areas of care where, on the production side, the ability to select providers currently exists as a theoretical but not necessarily a practical possibility for most citizens. In Sweden, which in 1991 made patient choice of site and health professional a central tenet of its health system, and in the

UK, which up until the 1991 reforms had officially allowed patient choice of general practitioner and, with a referral, official freedom to seek care at any hospital in the NHS, vouchers have certain advantages as long as other preconditions are satisfied. The degree to which patient choice is meaningful is also contingent on a variety of factors, including the availability of transportation and of adequate information about waiting times, and the quality and outcomes of the clinical services available (Saltman and von Otter 1992; Saltman 1992).

Potential problems

The potential disadvantages of the adoption of a voucher-style arrangement involve the same three actors, of course, but viewed from a different analytic perspective. A central problem for state funders is that vouchers confuse the distinction between equity of funding with equity of value received. In effect, there is the temptation of generating uniform vouchers for non-uniform groups of individuals. As advocates for the socially less well-off have maintained for decades, those groups require additional funding if they are not to be the subject of adverse selection but rather are to attain equal health status – that is, equal outcome (Culyer 1991). Recent evidence suggests that even in tax-funded public health systems, with no financial restrictions or reductions in access to services, levels of health inequality have increased during the 1980s (Whitehead 1988; Dahlgren 1992; Feinstein 1993). A related concern is whether vouchers disaggregate the interests of groups which, to ensure adequate social leverage, should be aggregated. It seems unlikely that a voucher programme based on individual distribution and control will adequately represent the collective interests of, say, the elderly as a group in the community (von Otter and Tengvald 1992). A third problem which vouchers present to national governments is the potential of regulatory capture. Large commercial providers in this as in other sectors of the economy could be expected to leverage their large volume of patient vouchers into a dominant voice in policy formulation, in a fashion that could readily compromise public accountability of decision-making (von Otter and Tengvald 1992).

A final but important issue is the potential loss of public accountability over the quality of services delivered. In areas of health care where quality is hard to monitor or evaluate, the temptation towards opportunistic behaviour (placing profit over patient interest) would increase (Williamson 1985). Furthermore, complaints about low quality could be readily dismissed if patients, by choosing the provider, had made themselves partly responsible for the outcome.

Although a source of potential difficulty for the state is that vouchers in health care could enable private interests to predominate over public interests, a source of uneasiness to service providers is, conversely, that public financial advantage might take precedence over the financial return to private providers. Specifically, the key problem for providers is that vouchers could require them to absorb financial risks involved in providing services to expensive populations. This underfunding could take many forms, including inadequate rates for high-risk populations and patients. A further risk is that politicians under financial pressure will seek to ratchet down the value of vouchers once the system is in place. For some providers, the risk exposure may be mainly in terms of time (i.e. for general practitioners under capitation), while for others it may be a direct financial exposure (insurance capitation). A related concern reflects the fear of new state regulation that could carry additional uncompensated and/or unrecoupable costs for service providers.

The greatest potential loser in a voucher scheme, however, may be the individual who is in need of services – the patient or client. If the voucher should be priced inadequately (intentionally or unintentionally), or if the service provider should reduce access or quality for various reasons (opportunistic behaviour, poor management), the patient may be left without necessary services. The disaggregated nature of vouchers requires this response to be at the individual level, yet vouchers almost by definition do not provide direct accountability of either service provider or public funder for the difficulties faced by the individual patient. If the patient has sufficient financial resources and the system allows, it may be possible to purchase additional services and/or supplemental insurance. The central dilemma for the patient, however, is that vouchers can encourage numerous opportunities for underservice, and that inadequacy in the delivery of services would become an officially sanctioned shift of cost onto the patient's private household resources – in effect, a shift of risk from the public funder to the individual in need of service. Experience in the Netherlands has demonstrated a different although related risk, that relations of dependent service recipients may try to take economic advantage of the situation, either by withholding benefits if they are transferable to cash or other goods, or by asking the provider to perform services for the household rather than the intended beneficiary.

Conversely, rather than subjecting patients to underservice, providers seeking to reduce costs and risk exposure could decide to accept vouchers only from more profitable patients – that is, to engage in adverse selection. This has been an endemic problem throughout both the private insurance and private service provider systems in the USA (Light 1992). Managed care providers like health maintenance organizations (HMOs) have sought to attract only healthy and/or 'young' elderly under the Medicare capitation programme by, for example, emphasizing their dental coverage for the elderly (high-risk elderly tend to have lost their teeth), and by insisting that potential enrollees must visit the centre during an open house (excluding the bed-ridden or seriously ill) (Luft and Miller 1988). In Sweden, the immediate result of the free right of establishment of private general practitioners has increased the imbalance in the number of physicians per capita between poor and well-off areas.

Private not-for-profit as well as for-profit hospitals in the USA have sought since the mid-1980s to 'improve their patient mix' by avoiding less well reimbursed Medicaid and other high-risk patients. In the UK, there has been considerable political debate as to whether GP fundholders have begun to employ adverse selection techniques in order to increase the profitability of their practices. Efforts in the Netherlands to prevent similar types of adverse selection, through the development of a sophisticated formula to risk-adjust annual premiums (i.e. each individual's voucher amounts), have thus far proven inadequate and unsatisfactory (van de Ven 1993; see de Roo, this volume). The upshot of this incentive for adverse selection by providers, of course, is that difficult, expensive and/or chronically ill patients – a category which disproportionately includes the poor – could lose out under a voucher system. In addition, the levels of compensation will be unfair to those personnel who do a socially responsible job as against those who choose the easy way out. This could in the long run lead to deteriorating services for the most needy groups.

EVALUATING VOUCHERS IN HEALTH CARE

The above exegesis suggests two separate conceptual frameworks through which to evaluate the likely impact of a voucher or voucher-based arrangement on a health care delivery system. The first is in terms of the ability to achieve the four governmental objectives noted above: micro-economic efficiency, budget predictability, individual choice and ideological change. The second reflects the relative consequences of shifting risks from public sector funders onto each of three major categories of health system actors: service providers, private insurers and individual patients. Taken together, these two frameworks provide a reasonable overview of the likely problems that a voucher programme would have to counteract if it were to be successfully implemented. We will review each framework in turn.

Achieving governmental objectives

A major attraction of voucher arrangements for governments is their potential to increase the *micro-economic efficiency* of both insurance provision and health services delivery. In each instance, the argument of proponents is similar: by splitting up an existing monopoly structure into competing sub-units, the cost of production will decrease as sub-units strive to increase their internal profitability, and the quality of services will increase as sub-units strive to attract subscribers/patients and thus expand their market share. A particular attraction of this market-oriented approach to constraining health care costs and increasing quality is that it can more readily fit programme priorities to the needs of the individual, as well as, ideally, facilitating lowered rates of taxation.

In evaluating the likelihood that vouchers can help improve micro-economic efficiency, a variety of factors must be considered. Current experience with attempts to introduce voucher-style arrangements for health insurance has already been detailed above. In the delivery of specific services, a commission looking into vouchers in Swedish municipalities has demonstrated relatively high transaction costs for setting up and administering voucher programmes. In Stockholm County, a bid to issue script vouchers (to the elderly for home care services) from a restaurant coupon company included a 2–6 per cent charge as the cost of printing the vouchers and administering their use.

Another voucher programme demonstrated the weak position of many clients at the point of delivery. Elderly and disabled patients in Stockholm, eligible for taxi services, found that drivers began to demand two vouchers for one trip. There were also allegations of vouchers being sold at a discount in a newly created black market – a process well known from food stamp programmes in the USA. In a voucher programme for dental services, there were stories that individuals with healthy teeth offered to trade their voucher to the dentist in return for a bottle of whisky.

Increased efficiency assumes that the voucher programme reflects the correct balance between integration and unbundling of services. In a Swedish experiment with vouchers for elderly care, separate vouchers were given for house cleaning, for cooking, for personal care, and so forth. This complete unbundling of services was demonstrably inefficient in that it procured and paid for services separately that, in a well-functioning home care system, typically overlap to make better use of the care-giver's time and – not incidentally – to reduce the number of care-givers involved and thus increase the likelihood of each care-giver feeling a personal responsibility towards his or her clients. Furthermore, complete unbundling ignores the reality that, on any given day, one or another patient or client might need more care, and that a well-designed system should be 'logistically' integrated to enable care-givers to balance several patients' need for time on a daily basis.

A third factor regarding the efficiency of voucher programmes concerns dynamic efficiencies – that is, reductions in cost achieved for delivering the same services. Among other dilemmas, a fixed price voucher per individual does not allow the issuing governmental agency to capture reduced per-unit costs. It will be in the interests of both the voucher holder and the producer not to allow gains from productivity to slip out of their control back to government. In the theoretical debate, this is sometimes remedied by allowing the holder to retain the value – at least in part – of unused coupons, and even trade them in for cash. This incentive for economizing might, however, easily lead to an unintended prioritization by the voucher holder.

This question of efficiency in per-unit terms raises additional issues. If, as in dentistry services, only a segment of the population needs intensive services, it may not be a good use of funds to design a voucher scheme for the entire population. Rather than being efficient, such a programme could perhaps instead be characterized as a wasteful bureaucratic exercise masked by a market-creating facade. Put more broadly, one could argue that collective contracts negotiated by public funders directly with providers may well be more cost-effective in numerous instances than individual voucher-style arrangements.

Moving from micro-economic efficiency to the issue of *budget predictability*, it is necessary to distinguish between short-, medium- and long-term outcomes. In the short term (i.e. one budget cycle), vouchers do appear able to cap expenditures effectively if all risks for unpredictable costs are shifted by the voucher to the provider and/or patient. In the medium term (several cycles), however, increases in the eligible population could well increase on-budget public expenditures. This is because vouchers stabilize the cost per individual, not the cost for the entire population entitled to services. Voucher arrangements nonetheless have the advantage of creating an effective point-of-control, rather than allowing a gradual in- crease of budget. In the long term, assuming increases in both service costs and the number of eligible individuals, vouchers may in a technically smooth way provide the cost savings necessary for budget stability, reducing the cost of each individual voucher by ratcheting down the value of the voucher in inflation-adjusted terms. This potential outcome could serve to shift additional finan- cial risks to service providers and recipients with less public atten- tion than in alternative arrangements.

An additional point concerning the cost of a voucher programme for health services concerns the need to establish an extensive and expensive programme of governmental regulation. To protect both taxpayers and the citizens' interest in the programme, health service providers participating in a voucher programme would need not only to meet explicit governmental standards, but also be subject to continuous monitoring and evaluation of their service quality and outcomes. Current Swedish, Finnish and British experience, sug- gests that such renewed regulatory oversight is essential to service quality once market mechanisms are utilized that create be- havioural incentives for underservice by providers (Saltman 1994). Even though providers and clients might create an 'unholy alliance', the long-term willingness of taxpayers to support these programmes depends on their socially beneficial results.

An assessment of vouchers in terms of the third governmental objective, *individual choice*, should reflect the broader structural context within which the choice decision is made. First, as noted above, in many publicly or privately financed health systems, patients already have a considerable measure of choice among general practitioners, primary health centres, hospitals and, to a somewhat lesser degree, specialist physicians. Moreover, patient choice can be tied to budgetary and/or payment mechanisms in a number of ways that do not involve the use of vouchers (Saltman and von Otter 1992). The addition of explicit vouchers – a paper chit

– would thus achieve nothing new either in facilitating choice or linking choice to budgets. Second, by placing a fixed price on the value of services an individual receives, a voucher system tends to restrict the range and potentially the appropriateness of services, and, for those services rendered, creates incentives for providers to select adversely, to reduce quality or to otherwise underserve. These incentives increase when competitive pressures weaken, as they most likely will once asset-specific investments (in a concrete and psychological sense) make the transition to a new provider difficult. Third, vouchers act to disempower patients, in that it is difficult for individual patients to hold voucher-paid providers – particularly privately capitalized providers – accountable for their service delivery decisions. The typical situation of an eligible client creates a weak negotiating position. Patients are often 'stuck' with their voucher-holder, who is insulated from public accountability by the voucher process itself (e.g. the public funder does not have line management authority over the provider, and the patient is seen as complicit in that he or she selected the provider him or herself). At one point in the earlier noted experiment with taxi vouchers for the elderly, for example, authorized drivers often chose to take fullfare-paying customers before the voucher-holders, leaving elderly individuals to wait in line.

For patient choice to be meaningful, as well as coherent with the public and social purpose of the programme, and to bring increased access to high-quality services in its wake, it should be exercised within a carefully constructed environment. The consumer function needs to be shared between the patient and the public planner, so that social and collective goods are reinforced (rather than diminished) by individuals' decisions, and to ensure that all offered choices are, in quality of care terms, choices that are 'good' for the patient (Saltman and von Otter 1992). To achieve this outcome, service providers should be required to meet explicit quality standards, and be accountable to public planning authorities through a continuous process of monitoring and evaluation.

Most importantly, there should be no artificially imposed financial cap on serving any individual patient, unless the risk basis is sufficiently well-known and individual outliers are not subordinated to abstract statistical constructions of a non-existent 'average client'. Such per-person caps encourage providers to lower quality, adversely select and otherwise underserve patients in need of care. A shared risk between public purchaser and provider for high-cost clients reduces the incentives both to under-serve as well as to push costs onto another party.

The ability of vouchers to achieve the fourth governmental objective, *ideological change*, is of course contingent on the policy goals of a sitting regime. Conceptually as well as empirically, however, vouchers seem directed towards dismantling the collectiv-ist use-value orientation of the traditional welfare state and replac-ing it with a possessive-individualist exchange-value approach to human services (Saltman and von Otter 1987). As a first step in a long-term process of privatization – both of risks and provision – vouchers appear to be attractive to certain political parties.

EVALUATING VOUCHERS AS RISK-ALLOCATORS IN HEALTH CARE

As already stated, vouchers can be assessed as an effort to reallo-cate risk from public funders (e.g. the welfare state) to one or another of three alternatives: to service producers, to insurers, and/or to patients and clients. Based upon the above conceptualiz-ation of the likely character, structure and impact that would (or has) accompanied the application of vouchers in health care, we can make the following observations.

Shifting risks to service producers

Similar types of risk reallocation already exist in many publicly operated health systems, in some circumstances and for some services. For example, diagnostic-related group (DRG) payments and other fixed-price systems for hospital care are utilized in a number of countries. Perhaps the key factor in making this risk-shifting acceptable is the degree to which savings in the delivery of short-term clinical services in fact reflects good management by the service provider, and the degree to which the incentives in vouch-ers, if properly constructed, encourage this efficient managerial behaviour. To satisfy social and political concerns, however, certain potential adverse consequences of this shifting of risk should be prevented. This suggests that the voucher process must be tightly ringfenced, through strict national standards and regulation, con-current monitoring and comprehensive periodic evaluation, to reduce the incidence of underservice, quality reduction, financial manipulation of provider accounts and adverse selection. Ideally, vouchers should be used for a condition which can be precisely defined and can be remedied by a specific service at a known price. In this situation, however, the need for a complicated voucher scheme would have disappeared. Thus, in reality, vouchers become appropriate for an intermediate category of cases where information is fairly transparent, but there remains a need for some discretionary judgement. Considered as a type of contract, therefore, vouchers ideally should not be completely 'hard', since they would no longer be needed, but neither should they be extremely 'soft', in which case transaction costs would increase.

To reduce the likelihood of unacceptable risk-shifting onto patients, given the differing characteristics and also varying openness to evaluation of different health sector services, the following additional caveats about issuing vouchers to service providers should be taken into account. The broader the categories of need that define a certain lump-sum group (same voucher value), the greater will be the difference, up or down, between the set value and the actual cost for handling the particular service(s). Thus, the rewards for adverse selection will increase. A reasonable overcompensation of costs for high-cost clients could compensate for this disadvantage and give vouchers a totally different connotation, although much of the cost advantage of the voucher system would disappear. Generally speaking, the socially responsible use of vouchers calls for adopting them for low- rather than high-intensity services, for simple rather than for more comprehensive services, for adults of sound body and mind rather than for individuals who are in a condition of mental or physical incompetence, for stable rather than highly variable needs, and for use in public or quasi-public rather than privately capitalized providers. None of these restrictions guarantee elimination of underservice or opportunistic behaviour, but, together with strict standards, monitoring and regulation, they serve to raise the threshold and reduce the likelihood that providers will engage in socially unacceptable behaviour.

Shifting risks to private insurers

Vouchers on the finance side of health systems appear to run considerably greater risks. The concept of risk-shifting from public funders to competing private insurers for the entire population, as has been proposed in the Netherlands, the USA and elsewhere, has a number of drawbacks (Saltman 1994). First, an insurance voucher for all medical services is too comprehensive and, at a one-year renewal period, is probably valid for too long a period of time. Both comprehensiveness and long validity invite private insurers to entertain opportunistic behaviour. Second, health insurance vouchers create powerful financial incentives for adverse selection and selective disenrolment of high-cost or high-risk patients. Third, it is difficult to monitor the behaviour of private insurers, or to regulate away the incentives towards adverse selection and/or underservice. Ten years experience with the behaviour of private insurers (not-for-profit as well as for-profit) in a competitive insurance market for managed care in the USA speaks eloquently on this point. Moreover, even if such regulation, monitoring and evaluation was technically feasible, it would be expensive. In essence, because the risks being shifted by vouchers on the finance side are so large, they create socially and politically severe consequences.

Vouchers on the finance side can also be expected to increase rather than decrease overall costs for health systems that currently have publicly run health insurance and/or a tax-based publicly operated health system. Overall transaction costs will be considerably higher. As *The Economist* – not always the greatest fan of governmental rather than market decisions – noted on this point (29 May 1993), 'government can pool risks and use its muscle to keep down costs better than any private' insurer can.

Shifting risks to patients/clients

The shifting of risk from a public funder to the individual patient is also problematic. The most common way to shift risk explicitly is to attach copayment or deductible schemes to a voucher programme (Nya Välfarden 1992). Such schemes typically are economically as well as clinically and socially counterproductive. Clinically, copayments encourage patients to self-diagnose, leading patients to visit a physician only for (often self-limiting) conditions that prevent them from going to work (e.g. flu) rather than to investigate more dangerous but less temporarily disabling conditions (e.g. having annual physical examinations) (Mechanic 1980). Copayments also encourage patients to postpone or avoid treatment, which can lead to more acute (and more expensive) care later on (D'Onofrio and Muller 1977). With regard to cost more generally, copayments have been shown to have no impact on overall costs of a health system; as Evans (1990) argues, the USA has the highest copayments yet also the highest expenditures. Moreover, copayments are an administratively expensive and inefficient way to collect additional health sector revenues (Evans 1990). Lastly, in social terms, copayments are regressive in nature, requiring disproportionately greater sacrifice from lower-income individuals, who often require more health services than better-off members of society. Two recent government-commissioned studies, countering claims by some health economists that copayments should be adopted to encourage the patient to be sensitive to the price of care (Pauly *et al.* 1991), concluded that copayments and deductibles could serve no useful purpose in a publicly financed health system (Richardson 1991; Evans *et al.* 1994).

CONCLUSIONS

Viewing vouchers from the perspective of designing planned markets, they appear to have the potential to improve the financial and/or the social performance of existing publicly operated health systems only if used in the specific intermediary circumstances outlined above. If one excluded capitation rates for primary care, and certain chronic conditions which need regular nursing but do not incapacitate the individual, one might conclude that voucher-type arrangements probably are not the most technically or economically preferred solution. Vouchers appear to have potential only on the production and delivery side of health systems. If they are to function in a socially and clinically appropriate manner, they require sophisticated external regulatory activity and expensive monitoring and evaluation over not only the separate provider institution but the administrative process of allocating vouchers as well.

Overall, the conceptual analysis above leads to the conclusion that vouchers can contribute to the construction of an efficient, effective and responsive planned market if applied in a carefully constrained manner. Vouchers should not be adopted, however, primarily as a device for ideological change without respect for essential welfare targets. Employed inappropriately, they can be expected to incur higher costs as well as resulting in lower levels of equity and overall health sector effectiveness.

NOTE

1 Research assistance for this chapter was provided by Shami Feinglass and Bill Haddock at Emory School of Public Health. Funding was provided by a grant from the Swedish Institute for Work Life Research, Stockholm.

REFERENCES

Culyer, A.J. (1991). 'Reforming health services: Frameworks for the Swedish review'. In *International Review of the Swedish Health Care System*, pp. 1–50. Stockholm: SNS.

Dahlgren, G. (1992). *Samhällets hälsopolitik 1992–2002 – Tänkbara framsteg och snedsteg*. Stockholm: Institutet för Framtidsstudier.

D'Onofrio, C. and Muller, P. (1977). 'Consumer problems with prepaid health plans in California'. *Public Health Reports*, 92, 121–34.

Evans, R. (1990). 'Tension, compression, and shear: Directions, stresses, and outcomes of health care cost control'. *Journal of Health Politics, Policy and Law*, 15, 101–20.

Evans, R.G., Barer, M.L., Stoddart, G.L. and Bhatia, V. (1994). *It's not the Money, it's the Principle: Why User Charges for some Services and not Others?* Toronto: Ontario Health Council.

Feinstein, J.S. (1993). 'The relationship between socioeconomic status and health: A review of the literature'. *Milbank Quarterly*, 71, 279–322.

Habermas, J. (1984–7). *The Theory of Communicative Action*. Vol. I–II. London: Polity Press.

Le Grand, J. (1990). *Quasi-Markets and Social Policy*. Studies in Decentralization and Quasi-Markets 1. Bristol: School of Advanced Urban Studies, University of Bristol.

Light, D.W. (1992). 'The practice and ethics of risk-related health insurance'. *Journal of the American Medical Association*, 267, 2503–8.

Luft, H.S. and Miller, R.H. (1988). 'Patient selection in a competitive health system'. *Health Affairs*, 7, 97–119.

Mechanic, D. (1980). *Future Issues in Health Care*. New York: Free Press.

Nya Välfarden (1992). *Eget Val i Äldreomvården*. Stockholm: MOU 1.

Pauly, M.V., Danzon, P., Feldstein, P. and Hoff, J. (1991). 'A plan for "responsible national health insurance"'. *Health Affairs*, 10, 5–25.

Richardson, J. (1991). *The Effects of Consumer Co-Payments in Medical Care*. National Health Strategy, Background Paper No. 5. Canberra: Department of Community Services and Health.

Saltman, R.B. (1992). *Patientmakt över vården/Patient Choice and Patient Empowerment: A Conceptual Analysis* (published simultaneously in Swedish and English). Stockholm: Swedish Centre for Business and Policy Studies.

Saltman, R.B. (1993). 'Public *vs* individual responsibility for health care'. In C. Artundo, C. Sakellarides and H. Vuori (eds), *Health Care Reform in Europe*, pp. 23–33. Copenhagen: WHO/Madrid: Spanish Ministry of Health and Consumer Affairs.

Saltman, R.B. (1994). 'A conceptual overview of recent health reforms'. *European Journal of Public Health*, 4, 287–93.

Saltman, R.B. and von Otter, C. (1987). 'Re-vitalizing public health care systems: A proposal for public competition in Sweden'. *Health Policy*, 7, 21–40.

Saltman, R.B. and von Otter, C. (1992). *Planned Markets and Public Competition: Strategic Reform in Northern European Health Systems*. Buckingham: Open University Press.

van de Ven, W. (1993). 'Regulated competition in health care: Lessons for Europe from the Dutch Demonstration Project!' Paper given at the *Annual Meeting of the European Health Care Management Association*, Warsaw, Poland, 29 June–2 July.

von Otter, C. and Tengvald, K. (1992). 'Vouchers: A revolution in social welfare?' *Economic and Industrial Democracy*, 13, 535–50.

Whitehead, M. (1988). *The Health Divide*. London: Pelican Books.

Williamson, O.E. (1985). *The Economic Institutions of Capitalism*. New York: Free Press.

7

CLINICAL AUTONOMY AND PLANNED MARKETS: THE BRITISH CASE

Stephen Harrison

INTRODUCTION

The principle of clinical freedom (the term 'clinical autonomy' is used synonymously) has been pervasive in the organization and management of the UK National Health Service (NHS). A working definition of this principle, whose hidden meanings we examine later, is that a fully qualified specialist physician or general practitioner is entitled to diagnose, treat and refer his or her patients as he or she wishes, within the limits of self-perceived competence and available resources. This is an elastic definition which essentially represents a claim to be unmanaged (Tolliday 1978), and has been treated as such by physicians; it has, for instance, been extended to allow surgeons to perform procedures of which they have little, if any, experience (Buck *et al.* 1987), to determine their own arrangements for timing clinics, and to determine the weight and type of their caseload (Harrison 1988).

In essence, therefore, clinical autonomy is a micro-level phenomenon, distinguishable in principle from the social and economic freedoms often enjoyed by physicians (Schulz and Harrison 1986). In the UK, these latter freedoms are to some extent the manifestation of macro-level sources of medical influence, exerted through close corporatist relationships between the British Medical Association (BMA), the physicians' trade union, and the Department of Health (Harrison *et al.* 1990: ch. 4). Ultimately, however, the micro-level practices are supported by the macro-level; for instance, it is national negotiating mechanisms that gave UK hospital specialists ('consultants') lifetime tenure and formal rights to make public comment, even in criticism of their employer. These rights exemplify successful claims to be unmanaged.

Planned markets in health care seem to threaten clinical freedom, both directly at the micro-level and indirectly at the macro-level. Thus, for instance, the notion of a purchasing agency which assesses aggregate population needs and priorities may well conflict with the needs and priorities of an individual patient as assessed by the clinician. Similarly, the notion of a purchasing agency seeking 'best value for money' among competing providers potentially undermines general practitioners' (GPs, i.e. primary care physicians) freedoms of specialist referral. Less directly, competition among providers of health care makes it imperative for managers to gain control of physicians; otherwise, the former cannot guarantee to 'deliver the goods' desired by the purchasing agency. The UK version of planned markets places in managerial hands a number of tools with which to effect such control.

This chapter has two main purposes. First, it shows how, and why, clinical freedom has been so pervasive throughout the history of the NHS; in particular, why it survived the challenge posed by the managerial reforms of the 1980s. Second, it examines the further challenges to clinical freedom which have followed from the recent introduction of planned markets to the NHS. This examination includes both the immediate consequences in terms of medical practice and the emergent political and ethical consequences.

The chapter is divided into four main sections. The first traces the history and practice of clinical autonomy from the inception of the NHS until the late 1980s, showing how it persisted in the face of developing managerialism. The next section analyses the politics of clinical autonomy in order to construct an explanation of its persistence. The third section gives an account of how the introduction of planned markets has further challenged clinical autonomy. The final section examines the unsought political consequences of this latest challenge – a public debate about health care rationing and a potential crisis for medical ethics.

Although there is no particular reason to believe that this analysis would not apply to other parts of the UK, the evidence here (as well as the organizational terminology) is from England.

A SHORT HISTORY OF CLINICAL FREEDOM

Clinical autonomy in policy and practice: 1948–83

It was noted above that clinical freedom has been pervasive in the NHS since its creation. This sub-section demonstrates this in three ways, two at the level of policy and one of practice.

Policy statements and clinical autonomy

In the early stages of the design of the NHS, clinical freedom was stressed as a key principle. Even before the end of the Second World War and the subsequent advent of the Labour government which established so many of the institutions of the British welfare state, the (coalition) government's white paper (policy document) on a national health service stated that: 'Whatever the organisation, the doctors taking part must remain free to direct their clinical knowledge and personal skill for the benefit of their patients in the way in which they feel to be best' (Ministry of Health 1944: 26).

After the war, Aneurin Bevan, the Labour Minister of Health who created the detailed institutional arrangements for the NHS, frequently reiterated these sentiments (Watkin 1975: 139; Allsop 1984: 17). The same view underpinned the first great reorganization of the NHS in 1974, which was initially planned by a Labour government, designed in detail by a Conservative government, and implemented by Labour. The Labour green paper (government consultative document) of 1970 said that 'the Service should provide full clinical freedom to the doctors working in it' (DHSS 1970: vi). Two years later, the Conservatives assured the reader that:

> The organisational changes will not affect the professional relationship between individual patients and individual professional workers [who will] retain their clinical freedom – governed as it is by the bounds of professional knowledge and ethics and by the resources that are available – to do as they think best for their patients.
>
> (DHSS 1972a: vii)

A more detailed government document from the same period began to make the link between clinical freedom and the managerial role:

> Success in [improving health care] depends primarily on the people in the health care professions who prevent, diagnose

and treat disease. Management plays only a subsidiary part, but the way in which the Service is organised . . . can help or hinder the people who play the primary part.

(DHSS 1972b: 9)

Three general elections later, the newly elected Conservative government under Mrs Thatcher stated in their white paper that:

It is doctors, dentists and nurses and their colleagues in the other health professions who provide the care and cure of patients and promote the health of the people. It is the purpose of management to support them in giving that service.

(DHSS and Welsh Office 1979: 1–2)

Formal organization and clinical autonomy

From 1948 until the mid-1980s, these principles were manifest in the formal organization structure of the NHS. A detailed account of such matters is inappropriate here (for such an account, see Harrison 1988: 9–22), but the central features can be summarized as follows. First, GPs continued to function as self-employed business persons under contract for services to the NHS. Managerial involvement with general practice was thus restricted to monitoring adherence to the (rather vague) terms of contract. General practitioners were entitled to refer patients to any specialist in any hospital anywhere in the UK, and (subject only to managerial intervention in the most extreme cases) to prescribe from the pharmacopaea in whatever quantities they chose.

Second, doctors (especially hospital specialists) dominated the management of the NHS, not in the sense of being formally responsible for it, but in the sense of having an entrenched and effective veto. Thus the statutory bodies within the NHS (such as regional hospital boards and hospital management committees) had large numbers of doctors in membership (Ham 1981) until the 1974 reorganization. After this date (until 1984), the NHS was managed by multidisciplinary consensus decision-making teams (for a review, see Harrison 1982). At the operational (district) level of organization, there were three doctors (two elected by their clinical colleagues) on a team of six; the consensus arrangement gave each an effective power of veto. Third, consultant contracts of employment were carefully insulated from managerial discretion, being held at the regional, rather than operational, level of organization, and including a number of unilaterally exercisable rights, including that of private practice.

At the national level, these organizational features were backed by a strong 'corporatist' relationship between the government health departments and the medical profession, operating especially in the area of medical 'manpower' (Harrison 1981; Harrison *et al*. 1990: ch. 4).

Practice: The NHS manager as 'diplomat'

Research into managerial practice before the mid-1980s clearly shows the spirit of the policies described above manifest in the daily operations of the NHS. Summarizing some twenty-five research studies conducted over this period, I chose the term 'diplomat' to characterize the NHS manager: someone whose role consisted of providing facilities for professionals, of smoothing out conflict and other 'organizational maintenance' tasks (Harrison 1988: ch. 3). More specifically, this system of diplomatic management had four constituents.

First, managers were not the most influential players in the NHS – doctors were. This is not meant to imply that individual clinicians had overwhelming authority, but rather that it was the (clinically free) decisions of individual doctors about which patients to accept and how to diagnose, refer and treat them that, in aggregate, defined the services of the NHS (Haywood and Alaszewski 1980; Schulz and Harrison 1983). Of course, resource constraints were recognized, but were seen as essentially temporary, especially at times of NHS growth.

Second, managers did not behave in the fashion of the proactive, goal-driven, textbook model. Rather, they *reacted* to problems passed to them by others in the organization, especially professionals (Haywood 1979; Stewart *et al*. 1980; Harrison *et al*. 1984). Third, and as a corollary of the second point, NHS managers were *producer-oriented* rather than consumer-oriented (Haywood 1979; Lee and Mills 1982; Thompson 1986).

Finally, the character of change in the NHS was incremental, in the sense that managerial attention was devoted to securing change at the margin, especially in deciding how to spend additional resources. The corollary was that there was little concern with review or evaluation of existing services, and little challenge to professional practice (Brown *et al*. 1975; Hunter 1980; Glennerster *et al*. 1983).

The Griffiths challenge to clinical autonomy

The mid-1980s saw a number of challenges of clinical autonomy, largely (though not entirely) emanating from the 1983 Griffiths Report (NHS Management Inquiry, 1983). Some of these were resisted, unsuccessfully, by the medical profession, whereas others were not.

The most visible change was the replacement, in 1984, of consensus teams by individual general managers. Although initially fiercely resisted by the BMA (Harrison 1994) as well as the nursing profession, it rapidly became regarded as a desirable development by all the groups of players within the NHS (Harrison *et al.* 1992), and a number of physicians themselves became general managers. The new general managers were employed on short-term contracts with performance review and individual bonuses which might have been expected to focus their attention on corporate objectives clarified through a recently introduced regional review process. Moreover, a system of 'performance indicators' was developed so as to facilitate national comparisons between hospitals and health districts; a number of these indicators were 'clinical' in the sense that they effectively measured doctors' performance in terms of such variables as length of stay and patient throughput.

The Griffiths Report also resulted, albeit over some time, in the introduction of a system of workload-related budgets for groups of doctors, initially referred to as 'management budgets', subsequently 'resource management' (Pollitt *et al.* 1988). These were clearly modelled on the practice of commercial organizations, and represented an attempted response to the earlier impossibility of controlling medical workload. The Griffiths Report also recommended that NHS managers should become more consumer-oriented, an injunction which many took to heart through the medium of consumer surveys. It should be noted that, in principle, such data offer a challenge to the medical monopoly of 'knowing what the patient wants and needs'.

One other attack on clinical freedom was resisted by the medical profession but introduced nevertheless in 1984. This took the form of restrictions on a number of items which had previously been available on NHS prescriptions.

The impact of Griffiths

Reviewing twenty-four published post-Griffiths research studies up to 1990, Harrison and colleagues concluded that, despite defeats over the *form* of the Griffiths innovations, the medical profession had experienced little, if any, resulting loss of autonomy. These findings correspond to the observation that, *so far as doctors were concerned*, the 'diplomat' role for managers had not changed much. This can be summarized with reference to the four elements of 'diplomacy' outlined above (for the full review, see Harrison *et al* 1992: ch. 4).

First, the new general managers did not generally function in a hierarchical relationship to doctors. Collusion and negotiation continued as the *modus operandi*, and doctors retained considerable ability (both overt and covert) to obstruct managers' aspirations (Pindar 1986; Strong and Robinson 1990; Pollit *et al*. 1991).

Second, despite a tendency to employ proactive language, managers continued to operate to a reactive agenda, including problems raised by doctors. There was a change, however; increased financial pressures within the NHS, coupled with increased intervention by the government health departments, constituted a significant alternative source of problems for managers, thereby reducing their capacity to respond to internal matters (Flynn 1988; Williamson 1990). Third, although as we have already noted a good deal of ostensibly consumer-oriented activity occurred, this was largely 'window dressing' and, at best, did not challenge *doctors'* behaviour (Stewart 1989; Strong and Robinson 1990).

Fourth, the new general managers clearly struggled to find the means to scrutinize and control change in the clinical arena. Resource management initiatives developed much more slowly than the Griffiths team had hoped, and encountered numerous technical difficulties, without being able to show much of an impact at the level of clinicians' behaviour, except perhaps to make the latter realize that 'creeping developments' (that is, where clinical judgements effectively introduce new services or more expensive ones) would be discovered by managers sooner rather than later (Pollitt *et al*. 1988; Buxton *et al*. 1989).

In summary, then, the medical profession's defeats occurred at the macro-level, rather than the micro-level. At least one of the macro-level defeats, the introduction of new prescribing restrictions, certainly represented a constriction in clinical autonomy, as did the tightening financial situation, but it cannot be said that the Griffiths and related changes lived up to their ostensible challenge to medicine. This does not mean that they are irrelevant to clinical freedom. More recent changes, such as the planned market, which *has* affected clinical freedom, could not have been introduced without the foundation of the earlier changes.

ANALYSING CLINICAL AUTONOMY

Discussions of clinical autonomy tend to be polarized; either it is seen as the means by which a doctor does his or her best for the patient, or else it is seen as a strategy by which doctors contrive an easy life for themselves. What is less often remarked upon is that these two positions are not in fact a polarity – they are not incompatible. Moreover, there is a third, equally compatible, position which will be discussed shortly. In this section, the politics of clinical freedom is analysed in order to show why it has proved so durable in face of the challenges described in the preceding section. The account is structured around three perspectives, each of which is drawn from a different sociological perspective on the concept of professionalism (Harrison and Pollitt 1994: ch. 1):

1 *Clinical freedom as medical ethics.* In any health system which relies upon third-party payment, a tension arises which is not present with out-of-pocket payment – a clash of interest between patient and third-party payer. The doctrine of clinical autonomy is the means of resolving this tension; it provides that, within broader or narrower limits, and subject to patient consent, the physician will be free to act in the *patient's* interests. Thus, although this particular view of clinical autonomy is espoused and articulated by the medical profession, often couched as an ethical statement, it is *patients* who would expect to benefit. It is thus an important political symbol for a national health service.

2 *Clinical freedom as professional dominance.* A consequence of autonomy is that it brings intellectual and material benefits to physicians. Depending upon the particular health system (for an international comparison, see Schulz and Harrison 1986), clinical autonomy may permit the choice of practice type and location, of clinical interest and treatment modes, of workload and of the logistics of working life. Evidence from the UK suggests that doctors do not have fixed views about the appropriate boundaries of clinical freedom (Harrison *et al.* 1984); rather, it functions as a rhetorical device, to be employed in the face of perceived threats. Indeed, this was exactly how the BMA responded to the Griffiths Report's general management proposals (Harrison 1994). It is

this view of clinical freedom which attracts the criticism of academics and managers, even though it may be in the interests of the latter. But it is primarily *doctors* who benefit from these aspects of autonomy.

3 *Clinical freedom as a rationing device.* Despite what has been said above, there are at least two important senses in which clinical freedom is not freedom at all. First, medical judgements are not just spontaneously produced by doctors; in whole or part they depend on socialization internalized during professional training. This includes what is sometimes referred to as the 'medical model': the view that ill health equals individual pathology, and that medical interventions are individual ones. Thus, the physician cannot react to his or her patient's asbestosis by issuing a prescription for factory closure. Second, resources are finite and the doctor cannot prescribe everything from which his or her patients and potential patients might benefit. In third-party payment systems, therefore, and despite the formal ethics, doctors *ration* health care, even where they perceive that the benefits of doing more for a patient might be slight (Harrison and Hunter 1994).

It is easy to see how these two factors work for the benefit of *government*. First, they confine the apparent causes of, and remedies for, ill health to politically safe ground; crudely, they do not challenge capitalism. Second, they make the process of rationing health care in a system which in theory provides comprehensive care politically invisible – reduced to individual, fragmented and unrecorded transactions between doctors and patients and/or their relatives. This process has been graphically described by two American students of the NHS:

> By various means, physicians . . . try to make the denial of care seem routine or optimal. Confronted by a person older than the prevailing unofficial age of cut-off for dialysis, the British GP tells the victim of chronic renal failure or his [*sic*] family that nothing can be done except to make the patient as comfortable as possible in the time remaining. The British nephrologist tells the family of a patient who is difficult to handle that dialysis would be painful and burdensome and that the patient would be more comfortable without it.
>
> (Aaron and Schwartz 1984: 101)

Something for everyone?

The above analysis makes clear why clinical freedom has been so persistent and pervasive in the NHS. Simply, there has been something in it for all the main players. Although the three perspectives outlined above are very different from each other, they are not contradictory. The continued practice of clinical autonomy has advantages for patients, doctors and government. Yet the newest set of NHS reforms have contrived to reduce clinical freedom in an unprecedented manner.

PLANNED MARKETS AND CLINICAL AUTONOMY IN THE NHS

This section begins with a brief description of the form of, and rationale for, planned markets in the NHS, as they derive from government policy initially promulgated in the white paper *Working for Patients* (Department of Health 1989; for a more elaborate account, see, respectively, Harrison *et al*. 1990: annexe; Harrison 1991). It then examines the main ways in which clinical autonomy is further challenged by these developments.

Planned markets in England

The form of planned market adopted in the UK is commonly referred to as the 'purchaser–provider split', terminology which emphasizes that it is built around *institutional* purchasing agencies, rather than a system in which the patient adopts the role of proactive consumer (Saltman *et al*. 1990). Thus, district health authorities (DHAs) are responsible for this purchasing role in respect of a geographically defined resident population, mostly in the range of 500,000–750,000 persons. For this purpose, DHAs receive (via the regional health authorities) central funds based on their population size, modified by weightings for age and sex structure, for mortality and for material deprivation. With such funds, DHAs may purchase health services from any provider, NHS or private, in any location to which it is deemed appropriate to expect patients to travel.

A parallel development to the DHAs' purchasing role is that of GP fundholding. Such GPs (who are volunteer applicants for this

status) receive funds directly from the regional health authority for
the purchase of certain specified (not all) secondary care services on
behalf of patients registered with the practice. These funds are, of
course, deducted from the financial allocation to the DHA where
the relevant patients reside, since they relate to services which that
DHA will *not* have to purchase. Otherwise, DHAs are in effect
financially responsible for the referral decisions of non-fundholding
GPs, a point to which we shall return in the next sub-section. In
principle, GP fundholders retain any unspent allocation for investment in their practice.

On the provider side of the new organization, the rapid development of NHS Trusts has occurred, so that by 1995 virtually the
whole of NHS hospital and community services will be provided by
organizations which, at their own request, will have achieved this
status. Trust status means that the organization has a more independent management than in the past (they are accountable to the
Department of Health rather than to regional or district health
authorities), and new freedoms over terms and conditions of employment and the retention of year-end financial surpluses. They
are, however, dependent for funds upon successfully entering into
contracts with DHAs and GP fundholders to treat patients, though
they may also treat private patients.

The implicit rationale for this form of the purchaser–provider
split seems to derive from 'New Right' theorizing about the behaviour of public sector bureaucrats (see, e.g. Niskanen 1971; for a
critique, see Dunleavy 1991). The essentials of this diagnosis are
that, in the absence of competition, public services will be over-
supplied in order to boost bureaucrats' career prospects and other
utilities; at the same time, there will be few incentives for productive efficiency. A purchaser–provider split on the lines described
above might therefore be expected to have two desired effects.
First, an institution with responsibilities only for purchasing, and
therefore to some extent insulated from 'shroud waving' and other
manifestations of provider pressure, should be able to make more
rational decisions about how to use finite resources to meet priori-
tized health care needs. Second, the organizational divorce of the
two functions allows an element of competition, since it removes
the previously existing guarantee that funds allocated for the care of
the population of a given locality will be spent within provider
institutions located there.

The organizational arrangements described above can be regarded as a 'planned market' in at least two senses. One is that the

DHAs, and perhaps to a lesser extent the GP fundholders, are assumed to plan for the needs of their populations or lists of patients, rather than simply respond to individual patient demands. This agency arrangement has sometimes been referred to as a 'quasi-market' (Le Grand 1990). The second sense is that the Secretary of State for Health and the NHS hierarchy retain a good deal of control over the service, and can thereby override the operations of the market so as to avoid politically undesirable outcomes. Thus, for the first year of the new system's operation, DHAs were discouraged from making major changes in the pattern of patient flow, and thereby threatening the financial position of providers. More recently, the government established an enquiry into hospital provision within London, and is making administrative decisions about its rationalization, rather than leaving hospital closures to follow naturally from the operations of the market.

New challenges to clinical autonomy

As with the mid-1980s developments in NHS managerialism, the very introduction of the purchaser–provider split represented a defeat for the medical profession at the macro-level. For instance, despite the development and spread of resource management having been agreed by the government and the BMA to have been subject to academic evaluation, a government decision to implement it generally was taken some time before the evaluation became available (in the event, the latter turned out to be somewhat unenthusiastic; see Buxton *et al.* 1991). Another example concerns the creation of Trusts themselves. Although provider institutions were supposed, when volunteering themselves for the new status, to ensure that there was a favourable consensus among senior staff (including consultants), evidence suggests that such consensus rarely existed in the case of the earliest Trusts (for a case study of how managers of one aspirant Trust contrived to ignore medical opinion, see Peck 1991).

As noted earlier, however, such defeats over *forms* of organization do not necessarily equate with reductions in clinical autonomy. Moreover, in research terms, it is too early to be conclusive – the evidence is sketchy. Nevertheless, there are indications that the character of NHS management is once again changing, a trend that can be considered under the four headings used earlier when summarizing research findings.

First, managers are now much more clearly 'in charge' of provider organizations. Thus, consultants' contracts of employment are now held by Trusts, which are increasingly free to vary their terms. For instance, a number have been reported to be offering fixed-term contracts rather than lifetime tenure, with other restrictions on private practice. Such contracts are also more specific than before in terms of detailing the times and locations of clinics, operating theatre sessions, and so on (Harrison *et al.* 1992). Moreover, Trusts are adopting forms of internal formal organization and budgets which attempt to integrate clinical doctors into mainstream management. All Trusts have a medical director, and many are adopting 'clinical directorates' as their predominant organizing principle. Such directorates are usually broadly speciality-based and headed by a physician acting in a part-time managerial capacity, often working with an associated lay general manager (Packwood *et al.* 1992). It is easy to see how the widespread adoption of resource management fits into such a model (Packwood *et al.* 1991).

The combination of these changes with the environment created by the purchaser–provider split has created a situation of far greater managerial control over the *clinical* activities of doctors than before. Thus hospital managers can demand changes in clinical services on the grounds that, without these, there will be a loss of business (Harrison and Wistow 1992). Many provider managers have also taken the view that 'cost and volume' contracts – that is, ones which specify that a finite number of cases of a certain type are to be treated for a specific price – are preferable to 'block' contracts, which specify type and volume of caseload only in the vaguest terms. It follows that what is specified externally for contract purposes must also be specified internally for management purposes, so that resource management has become a vehicle for controlling clinical behaviour so as to conform with the requirements of the contract. Finally, it should also be noted that the DHAs' liability to pay for the referrals of non-GP fundholders has led to the need to control 'extra-contractual referrals' – that is, those that do not conform with the contracts entered into by the DHA. In effect, GPs no longer have their former freedom of referral of patients to the consultant or location of their choice.

The second and third headings of change are management orientation in terms of proactivity/reactivity and producer/consumer problems, respectively. Managers have become more outwardly oriented than before in their efforts both to cope with the developing organizational changes and to satisfy the government's political imperative to present the NHS reforms as a success; however, the orientation is not markedly towards the consumer. Some managers have been proactive in trying to position their Trusts advantageously in the market, but there is some evidence which suggests that the political environment has worked against such behaviour being rewarding (Harrison *et al.* 1994). Moreover, the political salience given to the length of waiting times for hospital treatment seems on some occasions to have overridden doctors' clinical considerations about urgency of treatment.

Fourth, the introduction of the purchaser–provider split is beginning to have a significant effect on managerial orientation towards evaluation and performance. Participation in medical audit is now a government requirement of doctors, and contracts between purchasers and providers contain some kind of definition of the standards to which work is to be performed. Although rudimentary in the earliest days of the reforms, these provisions are rapidly becoming more sophisticated, and show signs of developing into protocols to control clinical behaviour.

In summary, unlike earlier managerial reforms in the NHS, the purchaser–provider split has begun to deliver a good deal of what it promised (or threatened, depending on the perspective) in terms of managerial inroads into clinical freedom. (Of course, it is unlikely that the latest changes could have been effected without the earlier ones, especially the introduction of general managers, though this is hindsight and should not be taken to assume foresight on the part of those who designed the changes.) Moreover, as might have been predicted from doctors' known attitudes towards clinical freedom, these inroads have not met widespread acceptance among clinicians. But that is not the end of the matter; as we shall see in the final section, there have been other consequences.

DECLINING CLINICAL AUTONOMY: THE CONSEQUENCES

Earlier in the chapter, the concept of clinical autonomy was examined from three distinct, though not incompatible perspectives. It was argued that it was the appeal of at least one of these perspectives to the NHS's major stakeholders that had sustained the level of clinical autonomy prevalent in the service from 1948 until, perhaps, 1990. Yet, as we have seen in the preceding section, this has now begun to break down.

One interpretation of what has happened is that, in the context of a government hostile to professionalism and to the public services and facing a massive budget deficit, the second of the three perspectives on clinical freedom – the notion of professional dominance employed as a means for making working life easy for doctors – became an obvious target for attack. Hence, all the managerial changes which have been outlined above constitute attempts to reduce professional dominance. However, if the practice of clinical freedom is attenuated from one perspective, it is attenuated also from the other two perspectives. It seems probable that, even in the relatively short term, some unsought consequences will follow. Let us take each of the remaining two perspectives separately.

The first remaining perspective, it will be recalled, represented clinical freedom as a basis for medical ethics: as a perceived guarantee that the doctor is able to act in the patient's interests rather than those of some third party. The purchaser–provider split seems likely to undermine this appearance in several ways. One example is that patients of GP fundholders will recognize that the GP may have a financial interest in decisions about referral or non-referral. Another is that what is presented by doctors as clinical judgement may come to be seen as social judgement. One example of this possibility is the refusal to perform coronary artery bypass surgery on smokers; although it is true that smoking does affect outcome, so, equally, do other factors such as obesity and diet, which are not discriminated against (Shiu 1993). Extra contractual referral (ECR) restrictions may also be visible to patients who wish to be referred to particular specialists or locations. Finally, the development of overt scoring systems of clinical priority (see, e.g. Giles 1993), which will result in some patients never being treated, clearly offend the notion that the patient's interests are paramount.

An obvious longer term consequence of loss of confidence in the ethical position of NHS doctors is that those who can afford to do so will increasingly be tempted towards private health care insurance, though it is clear that such a switch will hardly solve the problem; insurers are as keen to control health care costs as governments (see, e.g. Laing 1992). In either event, the standing of the medical profession may well decline as a result.

The final remaining perspective on clinical autonomy was as a politically unobtrusive method for the inevitable rationing of health care. The functionality of this approach to rationing is, of course, undermined if the process becomes explicit. But this is increasingly what is happening as a result of the purchaser–provider split.

Although most explicit rationing to date has been in respect of high profile but (in the context of the total NHS) marginal services such as cosmetic surgery, assisted conception and varicose vein surgery (Harrison and Wistow 1992), the problem is becoming more pervasive. Denied vascular surgery on the grounds that he smoked, a Wakefield man was reported to observe:

I have worked since I was 14 up until recently and paid a hell of a lot in taxes to the government, both in income taxes and on the 40 cigarettes a day I smoked. Surely it is not too much for me to ask to have an operation that might ease my pain in my old age and make me live a little longer.

(*Yorkshire Evening Post*, 26 August 1993, p. 1)

In addition, many hospitals have been either unwilling or unable (which, it is not clear) to spread their contracted caseload evenly throughout the financial year, resulting in much publicized restrictions on routine cases, and cancellations of patients already booked for admission.

Such explicit rationing may well undermine confidence in the NHS, or indeed the UK welfare state more generally. In the long run, this will not be a great problem for established physicians; they will always be needed, and will soon redevelop patterns of employment that free them from the unwelcome attentions of managers (Harrison and Pollitt 1994: ch. 6). There will also be those of a certain political persuasion who will welcome the destabilization of the NHS (see, e.g. Minford 1991). It is possible, however, that the NHS has been an important source of what analysts such as O'Connor (1973), Habermas (1976) and Offe (1984) term the 'legitimation' of the capitalist state in the UK. If this line of analysis if correct, then the decline of clinical autonomy may contribute to a more generalized crisis of government. It is, of course, difficult to be certain of either the concrete consequences of this or of the means by which governments might evade them. But if these consequences are not to include a further twist to the existing spiral of social disintegration and political alienation, it is clear that a new approach to the legitimization of health care rationing has to be found (Harrison and Hunter 1994).

REFERENCES

Aaron, H.J. and Schwartz, W.B. (1984). *The Painful Prescription: Rationing Hospital Care*. Washington, DC: Brookings Institution.

Allsop, J. (1984). *Health Policy and the National Health Service*. London: Longman.

Brown, R.G.S., Griffin, S. and Haywood, S.C. (1975). *New Bottles: Old Wine?* Hull: University of Hull Institute for Health Studies.

Buck, N., Devlin, B. and Lunn, J.N. (1987). *Report of a Confidential Enquiry into Perioperative Deaths*. London: Nuffield Provincial Hospitals Trust.

Buxton, M., Packwood, T. and Keen, J. (1989). *Resource Management: Process and Progress*. London: Department of Health.

Buxton, M., Packwood, T. and Keen, J. (1991). *Final Report of the Brunel University Evaluation of Resource Management*. HERG Research Report No. 10. Uxbridge: Brunel University Health Economics Research Group.

Department of Health (1989). *Working for Patients*. Cm 555. London: HMSO.

Department of Health and Social Security (1970). *The Future Structure of the National Health Service* (The Crossman Green Paper). London: HMSO.

Department of Health and Social Security (1972a). *National Health Service Reorganisation: England*. Cmnd. 5055. London: HMSO.

Department of Health and Social Security (1972b). *Management Arrangements for the Reorganised National Health Service*. London: HMSO.

Department of Health and Social Security and Welsh Office (1979). *Patients First: Consultative Paper on the Structure and Management of the National Health Service in England and Wales*. London: HMSO.

Dunleavy, P. (1991). *Democracy, Bureaucracy and Public Choice: Economic Explanations in Political Science*. London: Harvester Wheatsheaf.

Flynn, R. (1988). *Cutback Management in Health Services*. Salford: University of Salford, Department of Sociology and Anthropology.

Giles, S. (1993). 'Rationing scheme will exclude minor illnesses from NHS'. *Health Service Journal*, 26 August, p. 7.

Glennerster, H., Korman, N. and Marslen-Wilson, F. (1983). 'Plans and practice: The participants' views'. *Public Administration*, 61, 253–64.

Habermas, J. (1976). *Legitimation Crisis*. Oxford: Polity Press.

Ham, C.J. (1981). *Policy Making in the National Health Service*. London: Macmillan.

Harrison, S. (1981). 'The politics of health manpower'. In A.F. Long and G. Mercer (eds), *Manpower Planning in the National Health Service*. Farnborough: Gower Press.

Harrison, S. (1982). 'Consensus decision making in the National Health Service: A review'. *Journal of Management Studies*, 19, 377–94.

Harrison, S. (1988). *Managing the National Health Service: Shifting the Frontier?* London: Chapman and Hall.

Harrison, S. (1991). 'Provider markets in English health care: Incentives and prospects'. In J.R. Bengoa and D.J. Hunter (eds), *New Directions in Managing Health Care*, pp. 31–44. Copenhagen: World Health Organization.

Harrison, S. (1994). *Health Service Management in the 1980s: Policymaking on the Hoof?* Aldershot: Avebury.

Harrison, S. and Hunter, D.J. (1994). *Rationing Health Care: Options for Public Policy*. London: Institute of Public Policy Research.

Harrison, S. and Pollitt, C.J. (1994). *Controlling Health Professionals*. Buckingham: Open University Press.

Harrison, S. and Wistow, G. (1992). 'The purchaser/provider split in English health care: Towards explicit rationing?' *Policy and Politics*, 20, 123–30.

Harrison, S., Pohlman, C.E. and Mercer, G. (1984). 'Concepts of clinical freedom amongst English physicians'. Paper presented at the *EAPHSS Conference on Clinical Autonomy*, King's Fund Centre, 8 June.

Harrison, S., Hunter, D.J. and Pollitt, C.J. (1990). *The Dynamics of British Health Policy*. London: Unwin Hyman.

Harrison, S., Hunter, D.J., Marnoch, G. and Pollitt, C. (1992). *Just Managing: Power and Culture in the National Health Service*. London: Macmillan.

Harrison, S., Small, N. and Baker, M.R. (1994). 'The wrong kind of chaos? The early days of a National Health Service Hospital Trust'. *Public Money and Management*, 14, 39–46.

Haywood, S.C. (1979). 'Team management in the NHS: What is it all about?' *Health and Social Service Journal*, 5 October.

Haywood, S.C. and Alaszewski, A. (1980). *Crisis in the Health Service: The Politics of Management*. London: Croom Helm.

Hunter, D.J. (1980). *Coping with Uncertainty*. Letchworth: Research Studies Press.

Laing, W. (1992). *UK Private Specialists' Fees – Is the Price Right?* Eastleigh: Norwich Union Healthcare.

Lee, K. and Mills, A. (1982). *Policy-Making and Planning in the Health Sector*. London: Croom Helm.

Le Grand, J. (1990). *Quasi-Markets and Social Policy*. Bristol: University of Bristol School for Advanced Urban Studies.

Minford, P. (1991). *The Supply Side Revolution in Britain*. London: Institute for Economic Affairs.

Ministry of Health and Department of Health for Scotland (1944). *A National Health Service*. Cmnd. 6502. London: HMSO.

NHS Management Inquiry (1983). *Report*. London: HMSO.

Niskanen, W.A. (1971). *Bureaucracy and Representative Government*. Chicago, IL: Aldine-Atherton.

O'Connor, J. (1973). *The Fiscal Crisis of the State*. New York: St. Martin's Press.

Offe, C. (1984). *Contradictions of the Welfare State* (edited and translated by J. Keane). London: Hutchinson.

Packwood, T., Keen, J. and Buxton, M. (1991). *Hospitals in Transition: The Resource Management Experiment*. Milton Keynes: Open University Press.

Packwood, T., Keen, J. and Buxton, M. (1992). 'Process and structure: Resource management and the development of sub-unit organisational structure'. *Health Services Management Research*, 5, 66–76.

Peck, E. (1991). 'Power in the National Health Service: A case study of a unit considering NHS Trust status'. *Health Services Management Research*, 4, 120–30.

Pindar, K. (1986). 'The visible persuaders'. *Health Care Management*, 3–9.

Pollitt, C., Harrison, S., Hunter, D.J. and Marnoch, G. (1988). 'The reluctant managers: Clinicians and budgets in the NHS'. *Financial Accountability and Management*, 4, 213–33.

Pollitt, C., Harrison, S., Hunter, D.J. and Marnoch, G. (1991). 'General management in the NHS: The initial impact 1983–88'. *Public Administration*, 69, 61–84.

Saltman, R.B., Harrison, S. and von Otter, C. (1990). 'Competition and public funds'. In L.H.W. Paine (ed.), *National Association of Health Care Supplies Managers Members' Reference Book and Buyers' Guide*, pp. 64–6. London: Sterling.

Schulz, R.I. and Harrison, S. (1983). *Teams and Top Managers in the NHS: A Survey and a Strategy*. King's Fund Project Paper No. 41. London: King's Fund.

Schulz, R.I. and Harrison, S. (1986). 'Physician autonomy in the Federal Republic of Germany, Great Britain, and the United States'. *International Journal of Health Planning and Management*, 1, 335–55.

Shiu, M. (1993). 'Refusing to treat smokers is unethical and a dangerous precedent'. *British Medical Journal*, 306, 1048–9.

Stewart, R. (1989). 'Pressures and constraints on general management'. *Health Services Management Research*, 2(1), 32–7.

Stewart, R., Smith, P., Blake, J. and Wingate, P. (1980). *The District Administrator in the National Health Service*. London: King's Fund.

Stewart, R., Gabbay, J., Dopson, S., Smith, P. and Williams, D.T.E. (1987). *Managing with Doctors: Working Together?* Issue Study No. 5. Oxford: Templeton College.

Strong, P. and Robinson, J. (1990). *The NHS Under New Management*. Milton Keynes: Open University Press.

Thompson, D.J.C. (1986). *Coalition and Decision-Making Within Health Districts*. Research Report No. 23. Birmingham: University of Birmingham Health Services Management Centre.

Tolliday, H. (1978). 'Clinical autonomy'. In E. Jaques (ed.), *Health Services: Their Nature and Organization and the Role of Patients, Doctors, and the Health Professions*. London: Heinemann.

Watkin, B. (1975). *Documents on Health and Social Services: 1834 to the Present Day*. London: Methuen.

Williamson, P.J. (1990). *General Management in the Scottish Health Service*. Aberdeen: University of Aberdeen, Department of Community Medicine.

PART III

CONSTRUCTING ENTREPRENEURIAL PROVIDERS

8

SELF-GOVERNING TRUSTS AND GP FUNDHOLDERS: THE BRITISH EXPERIENCE[1]
Clive H. Smee

INTRODUCTION

The traditional strength of the UK National Health Service (NHS) has been its ability to provide universal access to good quality health care at a total cost that is low compared with most other health systems. The reforms that have been introduced recently aim to retain these advantages while introducing greater incentives for efficiency and greater responsiveness to patient needs. They also aim to improve the quality of care.

The major reforms in the secondary care sector were set out in the British Government's white paper *Working for Patients* (Department of Health 1989a) and became operational in 1991. The central idea in the reforms is the introduction of internal markets for hospital care. The old monolithic command and control system is to be replaced by institutional separation between purchasers of hospital care and providers, with relations between the two sides being determined by written contracts. Hospitals compete to win contracts and hospitals which attract more business also attract more income. The introduction of internal markets has required major supporting institutional changes. The most important of these are:

- A change in the role of district health authorities (DHAs), from organizers and providers of hospital care to purchasing agents, identifying the health needs of their local populations (average 300,000), deciding what services are required to meet those needs and then shopping around among potential suppliers before placing contracts.

- The opportunity for hospitals and other health services that were previously directly managed by DHAs to volunteer to become self-governing Trusts.
- The opportunity for large general practitioner (GP) practices to apply to become GP fundholders with their own budgets or practice funds, covering a defined range of hospital services.

The aim of this chapter is to chart the development of the latter two institutions – Trusts and GP fundholders – since the *Working for Patients* reforms became operational. Taking Trusts and GP fundholders in turn, the chapter focuses on five main questions: What was the rationale for the new institutions? What are the characteristics of the early Trusts and GP fundholders? How are Trusts and GP fundholders using their new freedoms? With what results? How may Trusts and GP fundholders develop?

From the beginning, application for Trust status and practice budgets has been voluntary. But ministers have not regarded either institutional change as 'experimental' in the sense of requiring careful evaluation before judgements could be made as to how far and fast the changes should go. They have been content to leave the pace of expansion of the new institutional arrangements to the interests and judgements of health service managers, clinicians and GPs. The Department of Health has, of course, been able to influence the pace of roll-out through the provision of financial incentives and the operation of selection criteria. However, the incentives in terms of direct subsidies have been very small and have been confined to assistance with start-up costs. The selection criteria were initially quite tight, particularly for Trusts, but have probably eased somewhat with successive waves of applicants, reflecting growing confidence that the new institutions will not abuse their new freedoms. Nevertheless, the expansion of both new institutional arrangements has greatly exceeded the expectations of all but the most optimistic of the reforms' authors.

Because the reforms were not seen as experiments, the government did not think it necessary to establish comprehensive formal evaluation procedures. There have been evaluation studies by outside bodies, but in general these have been hampered by the speed with which the new institutional reforms have taken off and the number of other changes in the health care system that have been proceeding in parallel. The Department of Health has its own information systems for monitoring and administration purposes, but these have had to be radically changed to adjust to the new issues thrown up by the internal market. Moreover, in line with the emphasis on decentralization, a conscious decision has been taken to collect less information at the centre. For these various reasons, information on the structure, conduct and performance of the new institutions is still partial.

NHS TRUSTS

Rationale

NHS Trusts are public sector organizations with a high degree of autonomy providing secondary and community health care services. When a hospital or other provider unit becomes a Trust, its assets are transferred from the Secretary of State or local health authority (LHA) to the ownership of the new Trustees while a debt of equivalent value is held by the Treasury. The Trusts are accountable directly to the NHS Management Executive through the executive's 'outposts'. The outposts monitor their financial performance and business planning but they are not 'managers' of Trusts. Prior to the introduction of Trusts, hospital and other secondary care services in Britain were provided through units directly managed by DHAs, which in turn were, in theory, under the direct line management of the Department of Health.

The new organizational arrangements represented by the term 'Trusts' are expected to lead to better performance, in terms of improved quality, efficiency and responsiveness to patients, through two main mechanisms. First, through devolution of management. As in other parts of the public sector, or indeed the private sector, giving local managers greater independence is expected to lead to decision-making that is based on a better understanding of local needs and is more responsive to local factor costs. Unshackled from the procedures, processes and pressures of higher tiers of management and with less political interference, Trust status should encourage innovation and stimulate staff commitment (Department of Health 1989a).

The second source of expected benefits is the exposure of Trust management to the sharper incentives of competitive markets. With hindsight, Trust status was a precondition for the introduction of an effective internal market in health services: providers had to be given autonomy from DHAs if districts were not to continue to face a conflict of interest between their responsibilities for purchasing health services for local populations and for employing the main suppliers of those health services. With the separation of purchasing and provision, Trusts must compete in the quality and value for money of the services they provide if they are to attract contracts and income.

The new organizational arrangements for the provision of health care services can therefore be seen as offering two principal advantages over the centralized command and control system they replaced: operational independence and market-type incentives. Depending on whether they have a management or economics background, academic commentators tend to emphasize only one of these features and to neglect the other. They may also impose tests on the reforms, such as whether the new arrangements meet the normal criteria for efficient markets, that are relevant to only one of the rationales for the new organizational arrangements. Arguably, Trusts should be able to out-perform the monolithic organizational arrangements they replace, whether they act as autonomous units of a single NHS 'firm' or whether they behave as competitive units in a health service 'industry'.

The health reforms announced in 1989 gave *all* NHS hospitals greater delegated powers. What additional freedoms do Trusts enjoy relative to provider units that remain under the direct management of DHAs? (The detailed regimes are set out in the Appendix.) The main freedoms relate to: personnel policies where Trusts are no longer bound by centrally negotiated pay and conditions of service; capital where Trusts are free to determine their own capital expenditure priorities and make cases for capital expenditure direct to the NHS Management Executive; absence of central controls in general, consequent on Trusts' exclusion from much central regulation; and the creation of their own management structures. How far Trusts have been able to make use of these new freedoms is an issue to which we turn later.

The 1989 reforms introduced significantly different financial arrangements for both Trusts and directly managed units. Most importantly, both types of unit have been cut off from direct funding from the Department of Health and both must earn their incomes through contracts with health authorities, GP fundholders and the private sector. There are also similarities in the financial duties faced by the two organizational forms: both are expected to break even, on an annual basis for directly managed units and 'taking one year with another' for Trusts; and both are expected to achieve a 6 per cent return on assets. The key differences in terms of incentives to improve performance are that Trusts may retain surpluses and are free to borrow within their agreed external financing limit (EFL). Directly managed units have neither right. However, a Trust's EFL is effectively a form of cash limit on total expenditure (capital and revenue). There is some limited scope to vary capital and revenue expenditure within the EFL and also to utilize unplanned surpluses to increase capital expenditure, although the latter can be difficult to generate within the tightly regulated pricing rules. Questions have therefore been raised as to whether, compared with directly managed units, Trusts face any greater incentives to improve performance.

Characteristics of Trusts

Application for Trust status has been voluntary. But Trusts have not been entirely able to select themselves. Candidates for Trust status have been required to exhibit some of the characteristics thought most likely to ensure success in the new organizational form. To be successful, units must be able to demonstrate:

- the benefit and improved quality of services that will flow from Trust status;
- the financial viability of the unit;
- the involvement of senior professional staff, particularly clinicians, in hospital management;
- that management has the skills and capacity to run the unit effectively.

The number of provider units able to meet these criteria has built up rapidly. The first wave of fifty-seven Trusts started operating on 1 April 1991 and accounted for only 12 per cent of the total number of beds in the NHS and 13.5 per cent of total NHS revenue expenditure. By the time the third wave started in April 1993, there were a total of 292 Trusts, accounting for 67 per cent of all expenditures and 55 per cent of all beds. A further 145 units applied for NHS Trust status in the fourth wave, which became operational as of April 1994. With these, some 95 per cent of hospital and community health service expenditure will be delivered through NHS Trusts.

Large acute and teaching hospitals applying for Trust status have attracted the most attention but, as Table 8.1 shows, even on the widest definition they account for barely half of the total number of Trusts. There are large numbers of Trusts providing community

Table 8.1   NHS Trusts: Characteristics of the first three waves

Type	First wave (April 1991)	Second wave (April 1992)	Third wave (April 1993)	Total
Teaching	9	6	7	22
Acute	15	22	49	76
Acute and community	4	4	1	9
Specialist acute	6	5	–	11
Community	6	15	31	52
Learning disabilities/ mental health	–	–	11	11
Priority care	7	13	8	28
Priority and community	–	13	5	18
Whole district	5	13	13	31
Ambulance	3	7	14	14
Total	55	98	139	292

services and care for the mentally ill and disabled. There are also more than a dozen ambulance services and a significant number of 'whole district' services, though the latter have begun to be officially discouraged. Although there are still major regional variations in the coverage of Trusts, it is clear that throughout the country they are accepted as *the* management model of service provision for the future. One consequence is that it is becoming increasingly difficult to compare the performance of Trusts with that of non-Trusts.

Conduct

The new organizational arrangements are expected to improve performance through the offer of operational independence and the introduction of market-type incentives. What signs are there that the expected changes in behaviour have occurred? Looking first at their new freedoms, it is clear (e.g. from the evidence submitted to the Health Committee of the House of Commons 1992–93) that most value freedom in the area of pay and personnel. But a survey of thirty-three first- and second-wave Trusts (one-fifth of the total) in February 1993 found that many Trusts have been slow to make use of this freedom. While most have set local terms and conditions of service, these generally applied to only a small proportion of the staff. Only a quarter reported comprehensive local pay and reward schemes and half of these were ambulance services. Lack of management time and lack of resources were cited as the main reasons for failure to introduce local terms of employment. A government ceiling of 1.5 per cent on public sector pay increases will also have discouraged action. However, the Trusts said they planned to get to grips with employment issues over the coming months. Numerous detailed studies of the utilization of hospital staff and the large variations shown in intra-hospital comparisons of labour productivity suggest that there remains substantial scope for the more efficient deployment of labour (see, e.g. the unpublished work of Judith Harper).

Turning to the new capital arrangements, there is anecdotal evidence that giving Trusts direct control over their capital expenditure has led to better and more innovative use being made of the available money. Some have also exploited the opportunity to use higher than planned surpluses to make additional capital expenditures within year. But there is no hard evidence that Trusts have been able to expand their asset bases faster than other provider units. Access to capital is now on the same terms for Trusts as for directly managed units: capital is allocated to those provider units submitting the strongest business cases for development without reference to their status. Trusts are, however, demonstrating more awareness of the importance of careful management of their capital estate, including the appraisal of new projects. This reflects both their greater understanding of the new market realities, particularly the need to earn a return on all capital and to meet capital charges, and the critical banker-type scrutiny of the 'outposts'.

The third area of new freedom is management structure. There has certainly been considerable experimentation in this area, with a large recruitment of new management skills and significant increases in management salaries. Making clinician involvement in hospital management a condition for Trust status has almost certainly accelerated clinician management of budgets. But anecdotal evidence on clinician involvement in, and commitment to, resource allocation decisions suggests there is still much to do.

Another condition for the granting of Trust status was the identification of improved services and patient benefits that could flow from Trust status. There is much evidence that individual Trusts are taking advantage of their freedoms to innovate and improve services. These changes cover a very wide range of activities, from patient hotels, through new forms of care for people with learning difficulties, to fuller utilization of hospital theatres. How many of these initiatives would have occurred in the absence of the new Trust freedoms is necessarily a matter of judgement.

How far Trusts are willing to take advantage of their new freedoms is, of course, partly a question of the sharpness of the incentives they face. The internal market has in theory fundamentally changed the incentives faced by all provider units. As noted earlier, the extension of Trust status has probably met one precondition for the establishment of an effective internal market, but Trust status in itself does not bring any substantial sharpening in incentive, except in so far as it reduces the willingness of the host DHA to maintain a 'cosy relationship' with its local hospital. The key question here, then, is whether the provider side of the internal market is showing the characteristics that economic theory suggests are likely to lead to greater efficiency, choice and responsiveness. Le Grand and Bartlett (1993) have suggested that these preconditions include a competitive market structure, good information concerning the costs and quality of the services being provided, modest transaction costs and actors motivated to respond to market signals. They conclude that for the provider side of the market in general, and for Trusts in particular, these conditions do appear to be in place, or at least within sight. The implication is it is reasonable to expect that Trust status is improving provider performance in important respects, even if it is difficult to capture that improved performance in hard data.

Performance

It is in fact very difficult to say anything conclusive about the performance of Trusts in terms of outcomes. Bartlett and Le Grand (1994) have spelt out the reasons. These include (1) the introduction of Trusts on a voluntary basis and their rapid expansion throughout the system, making it extremely difficult to identify matched comparators, and (2) the parallel introduction of a large number of other measures, including separation of purchasers and providers, greater devolution of management responsibility to directly managed units, and a major increase in health service resources, all of which make it very hard to identify the separate impact of Trust status. Against this background, it is probably prudent to lay out what information is available on the separate performance of Trusts and non-Trusts and to leave it to the reader to judge how far the differences go beyond association and how they might reasonably be attributed to causation.

Looking first at financial performance, most first- and second-wave Trusts met their financial duties in the first two years of operation. In 1991–92, only eight of the first-wave Trusts achieved a return of less than the target 6 per cent, and most were significantly above it (helped by an initial over-valuation of the asset base), giving an average return on capital employed of 9 per cent. In 1992–93, only eight achieved a return significantly less than 6 per cent, and the simple average across all 150 or so Trusts was 8 per cent. In terms of the requirement to break even, taking one year with another, most Trusts appear to have interpreted this as a requirement to break even every year. In 1991–92 only two Trusts failed to break even, and in 1992–93 there were only eight material failures. The third financial objective, to operate within external financing limits, has proved a little more troublesome. In 1991–92 nine Trusts failed to achieve their EFL target, and in 1992–93 two missed the target by more than £1 million. It is easy to explain away this performance in terms of the close central monitoring of Trusts and the tight controls on purchaser behaviour in the first two years of the reforms. But given the initial expectations of many of the reforms' critics and the major change of financial regime that Trust status involved, these results are encouraging.

Turning to hospital activity, statistics for 1991–92 and 1992–93 indicate that in the first two years Trusts have out-performed other hospitals in terms of increases in the numbers of patients treated. As Table 8.2A shows, this is true of both first- and second-wave Trusts. A distinguishing feature of first-wave Trusts is their high proportion of day cases (Table 8.2B). Many analysts see day case surgery as a particularly good indicator of the efficiency with which health services are organized. The success of the first-wave Trusts may of course reflect their greater initial efficiency (Bartlett and Le Grand 1994). But there is little evidence that the second wave of Trusts enjoyed an initial advantage in unit costs: their early performance is therefore particularly encouraging.

The Department of Health is working on the development of fairly crude indicators of total costs per unit of activity and labour costs per unit of activity that could be used to compare the performance of Trusts and directly managed units. Initial work shows very large variations across units and confirms that first-wave Trusts had lower unit costs than other provider units. But it does not yet allow

Table 8.2   General and acute activity growth in teh NHS, 1990–92 to 1992–93

A. Ordinary admissions + day cases (finished consultant episodes)

	Number (000s)			Growth	
	1990/91	*1991/92*	*1992/93*	*1990/91–1991/92*	*1991/92–1992/93*
First-wave Trusts	1263	1370	1436	8.5%	4.8%
Second-wave Trusts		1443	1526		5.8%
Directly managed units	5574	4528	4692	7.1%	3.6%
Special health authorities	104	107	116	2.9%	8.4%
Total	6941	7448	7770	7.3%	4.3%

B. Day case percentage (of all ordinary admissions + day cases)

	1990/91	*1991/92*	*1992/93*
First-wave Trusts	19.6	21.9	24.2
Second-wave Trusts		20.3	21.8
Directly managed units	17.6	20.3	22.7
Special health authorities	22.4	24.4	32.4
Total	18.0	20.6	23.0

any conclusions to be drawn about the trends in relative costs of Trusts in comparison with non-Trusts.

Moving on from indicators of activity and efficiency, there is some evidence that Trusts are proving more effective in meeting patient expectations. A key political concern in England is waiting times for inpatient treatment. Table 8.3A shows that over the eighteen months to March 1993, first-wave Trusts were more successful in reducing the numbers of patients waiting for treatment for over a year than were other hospitals. For second-wave Trusts, the available information covers only six months and it is perhaps too early to expect them to be doing significantly better than the remaining directly managed units. Information is also available on the total numbers waiting for inpatient treatment (Table 8.3B), and

Table 8.3 Number of patients waiting for treatment as inpatients or day cases: Figures for England at end of each six-month period

A. Percentage waiting over one year

	Sept 91	March 92	Sept 92	March 93
First-wave Trusts	15.9	7.1	7.4	4.8
All hospitals *not* first-wave Trusts	16.5	9.2	8.8	5.9
Second-wave Trusts	–	–	9.5	6.8
All hospitals *not* first- or second-wave Trusts	–	–	8.6	5.6
All hospitals	16.4	8.8	8.6	5.7

B. Total numbers waiting (000s)

	Sept 91	March 92	Sept 92	March 93
First-wave Trusts	179	170	176	184
All hospitals *not* first-wave Trusts	770	748	764	811
Second-wave Trusts	–	–	187	204
All hospitals *not* first- or second-wave Trusts	–	–	577	607
All hospitals	948	918	940	995

this again shows that over the eighteen months to March 1993, first-wave Trusts have put in a better performance than other hospitals. But preliminary information for the second-wave Trusts, relating only to the six months September 1992 to February 1993, suggests that they have not done so well.

Information is currently being collected on the performance of every provider unit in terms of a series of 'Patient's Charter' standards and guarantees that include: waiting times in outpatient clinics; waiting times for initial assessment in accident and emergency clinics; cancellation of operations; and waiting times for an ambulance service. Data on these areas of performance will be published in 1994.

The early guidance on the establishment of NHS Trusts suggested that purchasing authorities would wish to carry out regular surveys of patient satisfaction with Trust services. In 1992, a centrally organized survey of eight Trusts covered 900 patients who visited the hospitals before and after they became Trusts. Of these, 48 per cent said services had improved, 44 per cent said that was no change and only 8 per cent said services were worse (Department of Health 1993). There is no central comparative information on trends in patient complaints about Trusts and about directly managed units.

Costs

The establishment of Trusts has involved set-up costs. These have risen in line with the increasing number of units coming forward for Trust status: from £12.5 million in 1990–91 to £26.6 million in 1991–92 and £34.3 million in 1992–93. These sums have been necessary to fund the establishment of new payroll and finance departments, marketing activities, shadow Trust Boards and other infrastructure necessary for Trusts to exploit their new freedoms. Subsequent running costs are also significant: one unofficial estimate is that the incremental cost of running an acute Trust is some £500,000–£750,000 a year, costs which in many cases have traditionally been carried by health authorities and the centre (Newchurch & Co. 1993). But these costs are small in relation to either the budgets of the first three waves of Trusts (totalling over £11 billion) or in relation to the value of their assets (totalling over £12 billion). Trusts will, of course, have benefited from other expenditures on implementing the NHS reforms, including the development of information technology systems and the enhancement of personnel, finance and other key staff functions and training. Establishing and monitoring the Trusts has also generated administrative costs for the Department of Health, regional health authorities (RHAs) and the new 'outposts' of the Management Executive. Because of the complicated way in which costs have been shifted between organizational levels, it is too early to make firm judgements about the total recurrent administrative costs of organizing the provision of services through autonomous Trusts rather than through directly managed units.

The number of management staff initially taken on by Trusts has attracted some interest. But so long as Trusts face price-sensitive markets, they are expected to be discouraged from recruiting administrative staff that do not pay for themselves through improved efficiency.

Looking ahead

With well over 90 per cent of hospital and community health service expenditure delivered through NHS Trusts as at April 1994, it might be expected that we could look forward to a period of structural and behavioural stability. But the whole point of introducing market processes into the health care system is to facilitate adaptability and responsiveness to change. And, as one perceptive health service commentator has noted, the most likely scenario is 'a period of continuous and unpredictable change' stretching into the indefinite future (James 1993).

It is possible to speculate about how Trusts may evolve under three headings:

● How may the configuration of Trusts change?
● How will Trusts' operational freedoms evolve?
● How competitive will be the provider market?

A fourth question is the most critical of all:

● Will Trusts deliver the efficiency gains and service improvements expected of them by government and customers?

Each of these questions is considered briefly in turn.

The number and characteristics of the 440 or so Trusts that are expected to be in operation from 1995 largely reflects history. As market conditions evolve, there are bound to be pressures to change current configurations. As purchasers increase their effectiveness, there will be strong incentives for Trusts to merge – a process that has already been seen in the Netherlands. High costs and over-bedding in inner cities is another factor pushing in the same direction. Some mergers may follow the common pattern in the USA of central city hospitals amalgamating with those in the suburbs or dormitory towns. The critical issue will be whether the centre simply monitors and manages these changes or actively attempts to change the hospital structure in order to maintain or increase competition. The latter approach could involve deliberately breaking up larger units into smaller ones, prohibiting mergers which restrict choice, and using the allocation of capital funds to encourage Trusts to expand into each other's 'territory'.

The second issue, how Trust freedoms will evolve, breaks down into (1) whether those freedoms will expand or reduce over the coming years, and (2) how active Trusts will be in exploiting their freedoms. Many Trusts probably view the history of the last two and a half years as marked by a series of infringements on the freedoms they thought they were being given and continuous discouragement to exploit what freedoms they retained. This may be a one-sided view, but ministers and the centre are finding it difficult to reconcile devolved accountability with the demand for detailed monitoring created by parliamentary and media interest in operational issues. In consequence, the centre is drawn into a whole range of issues, from hospital catering standards to the freedom of speech of hospital staff that it once expected to leave to the discretion of local management. The dilemma is that without substantial operating freedom, Trust management cannot be expected to produce a better performance than the old directly managed units, but that with such freedom there is bound to be a diversity of behaviours and performance. The existence of outliers is then seen – by the press, auditors and politicians – as a cause for central regulation.

How far Trust managers exploit their freedoms will, as noted earlier, depend in part on the incentives they face. The search to reduce costs through, for example, more efficient labour utilization, will depend on what sticks and carrots are perceived. Many see the government's treatment of London as a key test of how far market processes will be allowed to go and whether hospital closures and the withdrawal of capacity are real options. (It is widely accepted that there is excess hospital capacity in London. At issue is whether government intervention will accelerate the adjustment process – the firm intention – or retard it, and how far the form taken by the downsizing will be guided by market signals.) But major closures are bound to be highly political events. More flexible financial rules aimed at facilitating small-scale downsizings could send equally potent messages. The obverse of making downsizings easier and therefore more feasible is that expansion and innovation should be encouraged. This would have implications for the treatment of surpluses, for pricing rules and for the allocation of capital funds. More broadly, given the local monopoly position of many Trusts, it is difficult to see how competitive pressures can be sustained without the active pursuit of pro-competitive policies from the centre. In the absence of such a development, Trusts' operational freedoms may become more constrained than those of the branches of national retail chains.

Whether over the long term Trusts can deliver the efficiency benefits and service improvements expected of them will depend in large part on what answers emerge to the questions posed earlier in this section. In particular, it will depend on how far and fast Trusts are able to exploit their freedoms over pay and staffing; how successful they are in involving clinicians in resource allocation decisions; how sharp the incentives are to improve technical efficiency, notably in relation to the retention and reinvestment of surpluses; whether natural tendencies towards the development of monopoly power are resisted by strong purchasers who are close to their customers; and whether the regulatory regime in which Trusts operate is clear and well defined, providing both hard budget constraints and swift exit paths for managements that do not perform and incentives to innovate and expand to managements that perform well. Many of these desiderata raise difficult trade-offs, for example, between meeting the expectations of local communities and meeting national objectives, and between technical efficiency and service innovation.

GENERAL PRACTITIONER FUNDHOLDERS

Rationale

General practitioner fundholders are self-employed GPs (primary care doctors) who manage a budget from which they are responsible for securing a defined range of hospital and primary care services for their patients. When a GP practice becomes a fundholder, budgets representing roughly one-fifth of the per capita costs of hospital and community health services are transferred from the DHA to the management of the practice. The transferred budgets cover: a defined group of surgical inpatient and day case treatments covering most elective procedures; outpatient services; diagnostic investigations; prescribing costs; certain practice staff; and, a recent addition, community nursing services. Fundholders are accountable to the NHS Management Executive through RHAs. Regions frequently delegate day-to-day management of fundholders to family health service authorities (FHSAs) who also have responsibility for other GPs and for primary care services in general. The details of the accountability arrangements are still being refined. The regions have only recently taken over responsibility for FHSAs and the system of reimbursement for GPs' core functions (as distinct from their budget-holding responsibilities) remains the subject of a national contract (the General Medical Service or GMS).

As set out in a 1989 working paper, the aims of the scheme are to:

offer GPs an opportunity to improve the quality of services on offer to patients, to stimulate hospitals to be more responsive to the needs of GPs and their patients and to develop their own practices for the benefit of their patients. It will also enable the practices which take part to play a more important role in the way in which NHS money is used to provide services for their patients.

(Department of Health 1989b)

Several mechanisms are seen as contributing to the achievement of these benefits. First, greater efficiency is expected to be achieved through giving GPs responsibility for the costs of the resources that their decisions effectively commit. Fundholders now have incentives to give more careful consideration to the cost-effectiveness of the drugs they prescribe and to weigh up the relative costs and benefits of hospital referrals as against drug-prescribing. Second, moving decision-making closer to patients is expected to lead to greater patient sensitivity. General practitioners in Britain have long acted as both gatekeepers to secondary care and as advocates for their patients. Armed with secondary and community care budgets their powers in both roles are substantially enhanced – they can now not only refer patients to other service providers, but can contract with those providers for the delivery of services of defined quality and timeliness. Fundholding represents a substantial shift in the balance of power from hospital consultants towards GPs and primary care.

A third source of expected benefits is the injection of competition into both sides of the 'internal market'. Before the spread of Trust status, fundholding forced hospitals and other providers to compete for funds from purchasers who – unlike health authorities – had no managerial responsibility for the continued viability of any particular group of providers. Of more continuing importance, GP fundholders also provide a yardstick with which other GPs and the media can measure the purchasing behaviour of DHAs. If that behaviour is seen to be deficient, the authorities face the threat that their power and influence (and their resources) will be drained as fundholding expands.

What incentives do GP fundholders face to take on these new roles? Practice budgets and GPs' personal incomes are, in theory, completely separate. But efficient and effective management of practice funds offers GPs three distinct benefits. These benefits will motivate different doctors to differing extents. The first benefit is quoted most frequently by fundholders themselves – that is, the opportunity to improve the quality of potential care through influencing provider behaviour and, in particular, through steering resources between different uses. The second and linked advantage is the ability to build up the range and quantity of services through generating surpluses. Compared with NHS Trusts, fundholders face far fewer restrictions on their ability to generate surpluses, to transfer them between years, and on how they may be invested (see Table 8.4). A third source of incentives is that practice funds can be managed to generate personal income. This can be achieved in two ways. First, GPs can spend their budgets on services which attract patients to the practice, raising capitation income; second, they can use their budgets to employ staff to undertake income-generating activities, such as immunizations, cervical cytology and health promotion.

The range of freedoms given to fundholders has led to concern that it may be difficult to prevent cream-skimming, cost-shifting and underprovision of care. The initial system of allocating budgets primarily on the basis of historical use of services limited the incentives to cream-skim. As there is a move towards formula-funding, these incentives could become greater, assuming that GPs are better able to predict patient costs than the formula setters. But the incentives are likely to remain muted because of the absence of a direct link between the provision of fund services and GP personal remuneration. Incentives to cost-shift and to underprovide are also dampened by the lack of a direct link between budget surpluses and GP remuneration. Moreover, the mechanisms through which fund-holding can be used to increase personal remuneration – by attracting more patients and providing services for which there are fee-for-service payments – discourage underprovision and cost-shifting. While the potential perversity of some of the incentives facing fundholders remains the subject of much speculation and some scrutiny by economists and others, there is so far little hard evidence of undesirable effects.

Characteristics of GP fundholders

Like applications for Trust status, applications for practice budgets have been voluntary. All fundholding units have had to be above a certain size – initially a patient list size of 9000 or more, subsequently revised to 7000 or more. But apart from this and a general injunction from the centre that fundholders have adequate managerial and computing capacity, the detailed selection criteria have been left to the RHAs. The regions have adopted systems of varying

Table 8.4   NHS Trusts and GP fundholders: Comparison of freedoms and incentives

Freedoms and incentives	Trusts	GP fundholders
Access to capital	Rationed through centrally determined EFL	Same as for self-employed in private sector
Generation and deployment of surpluses	Restricted by tight pricing policies on services sold to NHS	Cannot be used to increase GPs' personal incomes; otherwise no formal restrictions
Personal policies	Formal freedoms *initially* limited by inherited contracts and central medical manpower policies	No restrictions on terms and conditions of non-medical staff; however, posts require FHSA approval
Personal financial incentives	Performance-related pay for senior managers; no equity stake	Effective deployment of practice funds can generate increased personal income for GPs; self-employed principals also have equity stake in premises
Services	Free to determine range and extent of services subject to finding contractor	Within limits of scheme free to determine range and extent of services offered; no need to identity contractor
Accountability (a) to ministers	Each Trust Board directly accountable to Secretary of State via the NHSME	Fundholders accountable to regions for efficient and effective management of fund (and through them to Secretary of State)
(b) to patient/ local community	Publish annual report on performance, summary business plan and annual accounts; annual public meeting; community health councils may visit Trusts	Publish annual Patient's Charter statements on performance; patients have power of 'exit'

sophistication and with weights attached to different characteristics (see, e.g. Glennerster *et al.* 1992). But there is little doubt that the first waves of fundholders have been distinguished from their peers by lower aversion to the financial risk involved, managerial competence, the existence of well-established computerized data systems and perhaps particular ideological and political loyalties (Penhale *et al.* 1993).

As with Trusts, the number of volunteers has built up rapidly. The first wave of over 300 budget-holding practices that began operation on 1 April 1991 covered 7.5 per cent of GPs and a similar proportion of the population. The second wave almost doubled these figures and with the introduction of a third wave of fundholders in April 1993, there was a total of 1235 practices organized into 1120 funds incorporating some 6000 GPs and covering 25 per cent of the population. Over 800 practices in 600 or so Funds came into operation from April 1994, extending the coverage of the population to about one-third. With grouping arrangements and work currently being undertaken to facilitate entry to the scheme for smaller practices, the only effective constraints on the spread of fundholding are the willingness of GPs to join the scheme, and management capacity.

There remain major regional variations in the coverage of fundholders (see Table 8.5). Overall, rural and suburban areas are disproportionately over-represented and inner cities under-represented. The budget-setting process has thrown up major differences between funds in their historical use of services and hence in their capitation amounts. Contrary to much speculation, on average their initial capitation receipts have been set at a level below the national average spend on the services covered by fundholding budgets. This would appear to imply that before joining the scheme, the first waves of fundholders had been below-average users of hospital services and prescriptions. However, for referrals, one small-scale study found the reverse (Coulter *et al.* 1993).

Conduct

When the introduction of fundholding was being planned, the Department of Health gave some thought to how it might change GP behaviour. Reductions in prescribing and/or increases in generic prescribing, reductions in hospital referrals and/or increased use of private hospitals, and the development of a range of new

Table 8.5 Growth of fundholders by region

Regions	First and second waves			Third wave			Fourth wave (provisional)		
	Number of funds	Number of GPs	% Pop. covered	Number of funds	Number of GPs	Cum. % pop. covered	Number of funds	Number of GPs	Cum. % pop. covered
Northern	38	231	15.0	31	161	24.8	17	87	30
Yorkshire	55	228	18.9	53	249	32.7	24	108	38
Trent	54	316	13.8	84	420	31.3	73	314	44
East Anglia	13	85	8.0	35	193	25.2	18	101	34
NW Thames	43	235	14.2	45	223	26.7	33	153	36
NE Thames	23	107	7.1	25	122	13.9	25	100	20
SE Thames	29	147	9.2	48	227	22.2	81	388	43
SW Thames	38	225	15.8	28	135	24.9	52	257	42
Wessex	32	197	13.3	22	128	21.8	35	191	34
Oxford	46	279	21.9	28	152	33.2	32	159	40
South Western	38	244	11.0	31	170	18.7	33	171	25
West Midlands	58	410	12.6	73	357	26.3	104	511	46
Mersey	44	227	19.1	40	196	35.0	40	151	47
North Western	29	161	8.7	37	168	17.4	56	253	30
Total	540	3202	13	580	2901	25	623	2944	37

services based on the practice, were all identified as possible behavioural changes. What signs are there that the expected changes in behaviour have occurred? The answer is that because of the time taken to build up new information systems, the signs are not as strong as might have been expected. Overall, though, there seems little doubt that significant behaviour changes have occurred and are occurring. As with the evaluation of Trusts, these changes also reflect the large number of other reforms that were going on at the same time, including the introduction of a new contract for GPs. It is again particularly difficult to distinguish causality from association.

Bearing in mind these caveats, the limited national data and the more plentiful evidence from local surveys suggest the following:

1 In terms of prescribing, fundholding is associated with a lower volume of prescribing (Penhale *et al.* 1993), a more rapid increase in generic prescribing (Coulter 1993) and a lower rate of growth in total prescribing costs (Coulter 1993) relative to non-fundholding controls.

2 So far as referral behaviour is concerned, one detailed study of first-wave fundholders in the Oxford Region found no significant difference in the rate of growth of referrals or in the use of private providers relative to a set of control practices (Coulter *et al.* 1993). But these findings relate to the first year of fundholding only. National data indicate that for 1993–94, first- and second-wave fundholders are planning an increase in their general and acute inpatient hospital activity that is well below the rate planned by other purchasers (a growth of 1.9 per cent compared with 3.2 per cent).

3 The most widespread evidence of changes in behaviour relate to the development of new services. These centre on improving access for patients, both spatially by moving services nearer to patients, and temporally by reducing waiting lists. Widely reported innovations include consultant outreach clinics (i.e. consultants in GP surgeries) in areas such as diabetes and ophthalmology, testing in the practice (e.g. cholesterol, urine, pregnancy), often with equipment supplied by a local provider, and new services such as chiropody and physiotherapy (Corney 1993). Regional surveys indicate that the savings of fundholders are primarily being invested in premises and equipment in order to facilitate such innovations. These developments have not been confined to fundholders, but the present evidence is that they

have been on a larger scale in fundholders than elsewhere (Coulter 1993).

There is a considerable amount of evidence that GP fundholders have used their new budgets to inject competition into both sides of the internal market. These behaviours are not easily quantified, but there have been many reports of fundholders persuading hospitals to improve their services in the course of contract negotiation, if necessary through the threat of removal of their custom. The contracts negotiated by fundholders have generally been of a more sophisticated form than those initially used by DHAs and have often introduced quality standards that have subsequently been extended to all hospital patients. Further tributes to the energy with which these freedoms are used are the extent of complaints from hospitals that fundholders are 'destabilizing their finances' and from health authorities and other GPs that they are creating a 'two-tier service'. Sustained differences in services depending on accidents of physical location are obviously undesirable on equity grounds and rules have been drawn up aimed at minimizing inequities in treatment. But it is too often forgotten that innovation inevitably produces at least transitory inequity and that gains in services for some patients are not necessarily purchased at the expense of other patients. With the health service reforms having so many different components, judgements on the relative contributions of different elements are difficult, but overall fundholders appear to have done more than any other set of actors to interject market forces into the NHS.

Le Grand and Bartlett (1993) have noted that GP fundholders are particularly blessed with those characteristics that economic theory suggests are the preconditions of efficient markets. They have the information on patient needs and on the quality of many hospital and community services necessary to make them good agents. In most areas of the country, they face contestability even if patients are loath to change their doctor. As self-employed small businessmen, they have shown themselves highly responsive to market signals. Moreover, whether they are motivated by a desire to maximize their personal incomes, to promote the health of their patients, or to maximize the size of their budgets and responsibilities, the reforms have given them sharper and clearer incentives than any other group. Their one disadvantage is that because of the small size of their operations, they may generate high transaction costs – both for themselves and for providers.

Performance

For reasons similar to those already outlined for Trusts, it is difficult to say anything conclusive about the performance of GP fundholders in terms of outcomes. Once again we lay out what information is available on the separate performance of fundholders, other purchasers and other GPs and leave it to the reader to judge how far differences go beyond association and might reasonably be attributed to causation.

Taking financial performance first, there are no central figures on the percentage of fundholders who kept within their budgets in the first two years of the scheme. But aggregate savings amounted to around 4 per cent of 1991–92 budgets and were unlikely to be very different in 1992–93. Only two practices have left the scheme because of bad management, including financial mismanagement.

Because of the uncertainties involved in establishing initial budgets for GP fundholders, the generation of savings may not in itself be a good measure of efficiency. But there is other evidence suggesting that the existence of budget constraints is encouraging efficient purchasing. The best current evidence relates to prescribing expenditures. There has been an earlier reference to local studies suggesting that fundholding is leading to more economical prescribing. Table 8.6 shows the national picture. In 1991–92 fundholder drug budgets rose by 3 per cent less than the prescribing costs of non-fundholders, and in 1992–93 the difference was over four percentage points. But there is currently no central data on the growth of GP fundholder expenditures on non-drug items of their budget. (As noted earlier, there is some suggestion that they are expecting lower rates of increase in inpatient activity in 1993–94 than other purchasers.) There are also many individual reports of fundholders lowering costs through switching or threatening to switch referrals or other services such as pathology tests. In time, as budget-setting procedures become more formula-based, the best single measure of improvements in efficiency will be the annual saving on budgets.

What signs are there that fundholders are improving the quality of service on offer to patients? The importance of waiting times has already been referred to. There is plenty of local anecdotal evidence that fundholders have achieved reductions in waiting times. Regional surveys show a more mixed picture. The national data are still poor and must be treated as preliminary at best, but for what it is worth they do suggest that fundholders are achieving rather

Table 8.6 Increases in prescribing expenditure: Fundholders and non-fundholders 1990–91 to 1992–93

Regions	% Increase			
	1990/91–1991/92		*1991/92–1992/93*	
	Fundholders	*Non-fundholders*	*Fundholders*	*Non-fundholders*
Northern	14.0	13.9	7.6	11.4
Yorkshire	10.1	13.9	8.7	12.1
Trent	12.0	15.4	10.9	13.1
East Anglia	11.4	16.0	9.6	14.2
NW Thames	15.3	15.8	13.7	13.4
NE Thames	11.4	16.3	9.7	12.5
SE Thames	15.6	16.3	2.8	13.5
SW Thames	12.5	17.1	13.7	14.1
Wessex	10.9	14.7	6.9	12.8
Oxford	10.0	15.7	9.9	13.5
South Western	9.9	15.8	5.4	13.2
West Midlands	16.2	14.3	8.1	12.6
Mersey	9.7	15.3	3.8	13.4
North Western	9.8	13.5	8.8	11.1
ENGLAND	12.0	15.2	8.5	12.8

shorter waits than district health authorities. Inpatient waiting times are only one of the waits that causes dissatisfaction to patients. There is again anecdotal evidence that fundholders have reduced outpatient waiting times and have worked to reduce waiting times for GP appointments and waiting times in GP surgeries. A national database on waiting times for outpatient referrals is currently under development. No information is kept centrally on waiting times for GP appointments.

Other potential measures of responsiveness to patients and patient satisfaction are shifts in patient registrations, formal complaints and the results of patient surveys. There is anecdotal evidence that in some parts of the country fundholders are proving particularly successful in attracting new patients. There is no evidence at the centre on how complaints are split between fundholders and non-fundholders. Finally, all GPs are encouraged to survey their patients and it is known that some fundholders have followed this advice. But again there is no central information on how many fundholders regularly seek the views of their patients or on whether such surveys indicate greater satisfaction.

Costs

To assist with set-up costs, new fundholders have been given initial financial assistance. The management allowance for the preparatory year is currently £17,500 for a single practice fund and £20,000 for a group fund. The practice can also receive assistance towards computer costs of 100 per cent of the software costs incurred and 75 per cent of the hardware costs incurred. In subsequent years, fundholders receive support for the additional management associated with operating the fund. Reimbursement for ongoing administrative costs is against the actual costs incurred subject to a ceiling of £35,000. The management costs associated with fundholding work out at about 2 per cent of fundholders' budgets. For smaller practices, 'multi-funds' and agency models represent ways of holding down these costs.

Fundholding does of course give rise to other costs. There are administrative costs for RHAs and FHSAs in both the vetting of volunteers and in the subsequent monitoring of their financial and service performance. There are no central estimates of these costs. It has also been argued that the small scale and relatively sophisticated nature of fundholding contracts will add to the administrative and information costs of both providers and fundholders. In terms of information costs for providers, it is largely a question of bringing forward costs that would anyway have to be incurred as contracts become more sophisticated throughout the market. Because fundholders have more direct feedback on the outcomes and quality of different providers, they may require less expensive information systems than institutional purchasers. Critics have also referred to the additional costs of planning if district health authorities have to take account of GP fundholder priorities. But all health authorities are meant to consult their GPs, and to the extent that fundholders think more carefully about priorities and cost-effectiveness, they should be contributing superior information to the planning process (Glennerster *et al.* 1992).

Looking ahead

Although fundholding has expanded very rapidly, with a third of all primary care doctors having joined the scheme by 1994 there remains room for argument as to whether it will prove to be *the* model for the role of the primary care sector and the relationship between primary and secondary care. With an alternative purchasing model also in operation – the health authority – there is also room for debate about how purchasing will develop in the reformed NHS. Looking ahead it is possible to identify three important questions:

- How will the scheme evolve in terms of coverage of GPs and scope of budgets?
- How may the configuration of fundholders change?
- How will purchasing evolve in the internal market and what will be the relative roles of fundholders and health authorities?

There are two ways through which the role of fundholding could expand. First, more GPs could be brought into the scheme through allowing a reduction in the minimum list size or through encouraging small practices to share fund management costs. Second, through expanding the scope of the funds so that the services covered account for more than 20 per cent of hospital and community health services. The pressures to move in the first direction include the worry that otherwise primary care will become a two-tier service, with the quality of care available to patients depending on whether they live in proximity to large practices. However, the starkness of the difference can be reduced if health authorities offer non-fundholders real influence over their purchasing decisions. Success with current experiments in cooperative or joint management of funds could also end any perceived discrimination against small practices. Pressures to expand the scope of fundholding to cover a wider range of services are currently muted but may be expected to resurface once the more competent and energetic fundholders have adjusted to their current responsibilities. However, attempts to expand the scope of fundholding could run into conflict with attempts to extend coverage to smaller practices. In theory, the circle could be squared by allowing different forms of fundholding offering different ranges of services.

The current configuration of fundholding has primarily been determined by the interests of the initial volunteers. There are signs that the structure of fundholders may change in directions that could require central intervention if the benefits of the scheme are to be maintained. One such direction is the merging of purchasing activities into 'multi-funds'. There are already examples of such arrangements covering more than 100 GPs. The benefits of such a size are the same as the benefits prompting the merger of health authorities: greater market power and lower administrative costs. The potential dangers are restrictions on the individual practices' freedom and control over services for their patients, leading to loss of patient choice and responsiveness to patients. It is likely to be necessary to develop regulatory mechanisms able to weigh up these costs and benefits and armed, if necessary, with the power to strike down arrangements that unacceptably restrict patient choice.

A second way in which fundholding could evolve is away from being primarily a model for purchasing secondary care into being an integrated purchaser and provider similar to health maintenance organization. Stepping stones in this direction could include the use of surpluses to establish day care units and the development of relational contracts with 'preferred providers'. Developments of this kind are not necessarily to be discouraged, but again they suggest a growing role for a market regulator able to ensure that they are consistent with the maintenance of competition and choice.

The third question has been the most hotly debated and has been polarized around the issue of whether districts or fundholders should become the dominant purchasers. Supporters of the health authority solution tend to stress the pre-eminence of population needs and of meeting national priorities. Supporters of fundholding tend to give greater weight to individual patient demands and local responsiveness. In practice, ministers see no need to choose between the two models and the immediate issue is how they can most effectively live together. Do their roles need to be redefined to build on their relative strengths? What mechanisms will ensure that their actions are coordinated or at least consistent? These are questions that fundholders and health authorities are currently grappling with throughout the country. At present, sensible answers appear more likely to emerge from a process of natural experimentation than from the issuing of central dictats.

CONCLUSIONS

Working for Patients (Department of Health 1989a) saw NHS Trusts and GP fundholders as key organizational innovations in the new NHS structure. But initially they were not seen as central to the development of an internal market. In practice, both new organizational arrangements have spread more rapidly than was generally expected and both are now regarded as essential components of the internal market. Indeed, in the view of one senior health service manager, Trusts and GP fundholders are currently the only two 'fixed points' in the new structure of the NHS (Jarrald 1993). The wide adoption of Trust status came to be perceived as essential because it became clear that unless hospitals had separate management and accounting arrangements, it would be impossible to dismantle the local monopolies, conflicts of interest and cosy relationships that were a feature of the previous health authority structure. GP fundholders took on a critical role because they were fastest to demonstrate the benefits of a separate purchasing function and acted as a stimulant to the development of health authorities as effective purchasers. They have strengthened incentives on the purchaser role of the market through introducing an element of competition.

While there is no doubt about their importance for the development of an internal market, evidence of the contribution of both Trusts and fundholders to the fundamental objectives of the health service remains patchy and anecdotal. The reasons why hard information on outcomes is in short supply have been touched on earlier. New information systems that are currently being introduced should substantially improve the position, as will various ongoing research studies. However, it is in the nature of major organizational and management changes that they can rarely be 'proved' to be more effective than those they replace.

Few Trust managers and GP fundholders would wish to put the clock back and all political parties accept that self-governing hospitals are here to stay. But a subjective judgement would be that, relative to initial expectations, fundholders have so far proved more effective and certainly more conspicuously innovative than Trusts. Possible reasons for this include:

1 GP fundholders have in general been given greater freedom and stronger financial incentives than NHS Trusts (see Table 4). They may also have been subject to fewer central controls and a more stable regulatory regime.
2 As self-employed small businessmen, GPs may have had more of the skills required to work in an internal market than hospital managers and have found it easier to adjust to the new culture.
3 Fundholding has been established at a time when there is a worldwide trend to shift services out of hospitals into primary and community care settings. Fundholders have found they face an expanding market for their services as providers: technical

developments and public preferences both appear to give legitimate grounds for using the new funds to move services closer to patients and often into fundholder premises.

4 By contrast, acute and community Trusts have been born into a world of overcapacity and long-term decline in hospital beds. The contracting market they face has been made more transparent by the separation of purchasing and provision. Not surprisingly, many Trust managements have been putting their energies into fighting for their existence rather than into developing innovative and high-profile new services.

Other countries may wish to bear these points in mind when considering what lessons may be learnt from the UK's experience.

NOTE

1 The views presented here are those of the author only and should not be taken as representing those of the Department of Health or the government.

REFERENCES

Bartlett, W. and Le Grand, J. (1994). 'The performance of trusts'. In R. Robinson and J. Le Grand (eds), *Evaluating the NHS Reforms*. London: King's Fund.

Corney, R. (1993). 'Fundholding in general practice in South East Thames Regional Health Authority'. Unpublished manuscript, March.

Coulter, A. (1993). 'Evaluation of GP fundholding in Oxford RHA'. Unpublished manuscript, University of Oxford, March.

Coulter, A. *et al*. (1993). 'Effect of NHS reforms on general practitioners' referral patterns'. *British Medical Journal*, 306, 433–7.

Department of Health (1989a). *Working for Patients*. Cm 555. London: HMSO.

Department of Health (1989b). *Practice Budgets for General Medical Practitioners*. Working Paper 3. London: HMSO.

Department of Health (1993). *Government Response to First Report from the Health Committee 1992–93*. Cm 2152. London: HMSO.

Glennerster, H. *et al*. (1992). *A Foothold for Fundholding*. London: King's Fund.

Health Committee of the House of Commons (1992–93). *NHS Trusts: Interim Conclusions and Proposals for Future Inquiries*. First Report. London: HMSO.

James, J. (1993). 'Reshaping the NHS: From radical change to continuous change'. In *Health Care UK 1992/93*. London: King's Fund.

Jarrald, K. (1993). 'Service and structure'. Speech to the NAHAT Conference, June.

Le Grand, J. and Bartlett, W. (1993). *Quasi-Markets and Social Policy*. London: Macmillan.

Newchurch & Co. (1993). 'Strategic change in the NHS: 1. Unleashing the market'. Newchurch Health Briefing, August.

Penhale, D. *et al.* (1993). 'A comparison of first wave fundholding and non-fundholding practices'. Unpublished manuscript, City University.

APPENDIX

Comparison of the regimes of NHS Trusts and directly managed units

Issue	*NHS Trust*	*Directly managed unit*
Management	Each Trust is run by its own Board of Directors. The Trust is free to determine its own management structure. Senior professional staff must be involved in management	The DHA is responsible for the unit with the unit general manager (UGM) being responsible for day-to-day management. Internal management arrangements are subject to DHA approval
Accountability	Each Trust Board is directly accountable to the Secretary of State via the NHSME	Each unit is accountable to its managhing DHA. The DHA is accountable to the RHA and the RHA is accountable to the Secretary of State via the NHSME
Funding	Each Trust's income is largely derived from contracts with health authorities, GP fundholders and the private sector	Each unit's income is largely derived from contracts with health authorities, GP fundholders and the private sector
Services	Each Trust is free to determine the range and extent of services it wishes to provide, except that where a service must be provided locally a Trust can be obliged to provide it if it is the only unit able to do so	The range and extent of services offered by a unit are determined by the managing DHA.

Issue	NHS Trust	Directly managed unit
	Each Trust provides the services which it is contracted to provide	Each unit provides the services which it is contracted to provide
	There is no requirement for a Trust to consult the Community Health Council on closures or changes in use	Closures and changes of use are subject to formal consultation with the Community Health Council
Employment of staff	Each Trust sets its own staffing structure and levels. It employs all its own staff, including consultants	Each unit determines its own staffing structure, but its staff are employed by the DHA or RHA
	Each Trust is free to determine the pay and other terms and conditions of employment to *all* staff it employs. Staff transferring to a Trust retain their existing terms and conditions of service until changes are negotiated	Pay and other terms and conditions of employment of nearly all staff are subject to Review Body or national Whitley Council agreements or departmental determination
Financial duties	Each Trust has a statutory duty to break even taking one year with another. It is also required to achieve a 6 per cent return on assets and keep within its agreed EFL	The managing DHA has a statutory duty to balance its budget each year. The UGM has a managerial imperative to ensure that the unit breaks even. The unit has to pay capital charges equivalent to depreciation and 6 per cent interest on its fixed assets
Prices	Each Trust prices to cover running costs, depreciation and return on assets	Each unit prices to cover running costs and capital charges (depreciation and interest)

Issue	NHS Trust	Directly managed unit
Surpluses	Trusts may retain surpluses	Units cannot retain surpluses
Borrowing	Each Trust is free to borrow within its agreed EFL	Units have no power to borrow
Insurance	Trusts do not insure for clinical negligence. They may insure for other insurable risks	Units do not insure for clinical negligance or other insurable risks, with certain limited exceptions
Ownership of assets	Each Trust owns its assets. It is generally free to acquire and dispose of assets and retains the proceeds from any sales	Each unit's assets are owned by the Secretary of State or the health authority. Acquisition and disposal of assets are subject to control and regulation by the Department of Health, the RHA and the managing DHA. Retention of proceeds from sale is subject to decisions by the RHA
Capital	Each Trust makes its case for capital development to the NHSME. It funds the agreed programme from its own resources or by borrowing within its agreed EFL	Each unit makes its case for capital developments to the DHA or RHA. Funding is dependent on allocations from the regional capital programme

Source: NHSME Working for Patients (1990). Annex A of *NHS Trusts: A Working Guide*. London: HMSO.

IMPLEMENTING PLANNED MARKETS IN HEALTH SERVICES: THE SWEDISH CASE

Anders Anell

BACKGROUND

This chapter discusses the current reform process in the Swedish health care system, in particular the prerequisites for implementation of proposed market-oriented mechanisms. Empirical data were primarily collected in two studies (Anell and Svarvar 1993, 1994).

The financing and organization of Swedish health care is predominantly the responsibility of twenty-six county councils (including three independent municipalities). The county councils levy a proportional income tax on the population, and are authorized by the 1982 Swedish Health and Medical Care Act to structure most aspects of health service delivery. Nearly all hospitals in Sweden are owned and operated by the (elected) county councils, and most primary care is delivered through county-run health centres. Most county councils are responsible for providing care to a population of between 150,000 and 550,000 individuals. The exception is the County Council of Stockholm, which has a population of 1.7 million individuals.

Rapid growth and diversification of services have characterized the post-war development of the county council health care system. General growth of the Swedish economy made it possible to finance an expansion in the supply and accessibility of health services. This expansion was also marked by the transfer of responsibilities to the county councils from the private sector (e.g. outpatient care in hospital) and from the state (e.g. psychiatric care and teaching hospitals) during the same period. By the early 1980s, the county councils had in principle assumed total responsibility for the health care of their respective catchment areas.

During the 1980s, many counties encountered financial difficulties due to a general decline in economic growth combined with political pressures against increasing the county tax further. The importance of national government grants to the counties also decreased. Queues for elective surgery and other treatments lengthened. By the early 1990s, the county councils were being severely criticized from many quarters. The queues were discussed in the mass media, as were the county councils' formal and bureaucratic attitude towards patients. Previous studies of the productivity of the health care sector had shown substantial deficiencies (Lindgren and Roos 1985) and new case studies pointed to considerable potential for improved productivity in several spheres. A government report indicated the health care sector as one of the public services in which citizens felt most powerless (Petersson *et al.* 1989). Adding to the debate, the Swedish Medical Association called for the abolition of the county council-based health system.

Confronted by these problems, several county councils initiated major structural reforms. These included separating purchasers from health care providers, and promoting competition among providers. Freedom of choice for the patient, in which the money follows the patient's choice, was also an important political issue. In the past, counties have exercised their responsibility through a set of centralized planning and control systems. Politicians and senior management sought to influence health sector activities through plans, budgets and instructions.

REFORMS IN THE SWEDISH HEALTH CARE SECTOR

The following trends serve as the basis for reform of the Swedish county councils:

● patients are entitled to choose their primary care and hospital providers;
● elected officials should act as purchasers of services;

- responsibility for the provision of services should be delegated to providers;
- payment to providers should be performance-based rather than based on global budgets;
- an increasing part of service delivery should be subject to competitive pressures.

Although most health care politicians appear to agree that the health care sector needs to be reformed, they do not always agree on what the reforms should be. Approximately ten county councils have introduced changes based on some sort of planned market, and in several of these the changes are already significant. These changes usually comprise extended options in choosing health care providers, the establishment of public purchasing boards, and a performance-based system for paying hospitals, health centres and other health care providers.

The following analysis is based on the principles and experience of those county councils which are furthest along in implementing new control systems. These counties can be regarded as representative in that they reflect the leading edge of reforms in the Swedish health care sector. However, the selected counties are not representative in view of the overall Swedish experience of planned markets in the health care sector, which is rather poor. In several county councils, reforms are non-existent or of minor importance. It should also be remembered that reforms within the most advanced counties have only been in place for a few years.

Freedom of choice for consumers

Both at the national and county council levels, the issue of individual choice of health care providers has been given priority by the health care politicians. There does not seem to be any political disagreement about this issue. At the 1989 Congress of the Federation of Swedish County Councils, it was agreed by all political parties that freedom of choice should be strengthened. At the 1991 Congress, the decision to introduce freedom of choice was reinforced. A parliamentary committee appointed by the government in 1992 (HSU 2000, 'The Committee on Future Funding and Organization of Swedish Health Services and Medical Care') was given instructions that emphasized the freedom to choose one's provider of health care.

In the last few years, all county councils have increased the nominal options for individuals in that they may freely choose between primary health care centres and, in some cases upon referral, hospitals. In many cases, it is also possible to choose from providers in other counties. It should be noted that all Swedish hospitals have large outpatient departments, which adds to the array of possible alternatives for patients. In principle, there is a great difference compared with the previous system when a patient had to visit the primary health centre and hospital to which he or she was assigned (the Swedish term was 'to which one belonged'). However, the overall effects of freedom of choice have only been significant in the urban parts of the country. In the more rural areas, people tend to 'choose' the nearest primary health centre or hospital.

Many county councils also report changing attitudes towards patients among physicians and nursing staff. Earlier, there was much talk about putting the patient 'in focus'; nowadays, providers are compelled to do so. The reason lies in the financial incentives – the money shall follow the choice of the patient. The threat that the patient – and the money – might go somewhere else has often been enough to bring about a change in attitude. This effect corresponds well with the theory of contestable markets, in which it is the threat that somebody else could take over the activity, rather than the number of actual choices, that influences the response of the providers (Baumol *et al.* 1982).

In western Sweden (the area around the city of Gothenburg), there is far-reaching cooperation between five county councils on patient freedom of choice. Reimbursements are settled according to charges fixed in advance for inpatient (using diagnostic-related groups of DRGs) and outpatient cases. Freedom of choice in this region has created winners and losers among the county councils, since the money follows the choice of the individual. Individuals are increasingly seeking health services in the city of Gothenburg. The losers thus far appear to be two county councils which border on the city. This shift in patient patterns reflects in part the fact that county boundaries were fixed over a hundred years ago, prior to present-day work and travel patterns.

The flow of patients has also led to a change in attitude and to new efforts on the part of health care providers in and around Gothenburg. The purpose of the new efforts is to direct citizens' attention to the hospital in the county in which they live and to improve accessibility and quality. Several hospitals have undertaken marketing activities and, according to some observers, a passion for medical engineering has flourished – new techniques and good equipment look attractive in both information leaflets and newspapers. Various small and anecdotal changes also point to changed provider attitudes. In one hospital, it is now the patients who park their cars close to the hospital, not members of staff.

The flow of patients in western Sweden led to a controversy regarding how payments across county boundaries should be calculated. The financial effects were greater than expected and the situation in one county had already become very difficult. The result was that in 1993 the county councils obtained a partial refund on amounts paid, to the annoyance of the Gothenburg hospitals which had provided the care. At the same time, the rules of the game were changed. For 1994 and 1995, norms regarding the size of acceptable flows of patients were decided, and only 30 per cent of the normal reimbursement was paid for flows in excess of these norms.

In Stockholm, freedom of choice is in principle as far-reaching as that in western Sweden. However, the effects there are not as great and the nature of the problems that accompany patients shifting between providers is not the same. This may reflect the fact that these hospitals belong to the same county council. Thus, it is not a matter of paying money to other county councils as a result of patient choice. Just as in western Sweden, some Stockholm hospitals lost patients due to the fact that catchment boundaries did not correspond with the population's preferences. In some instances, freedom of choice made it possible for some people to choose the hospital nearest to their home.

There are instances which suggest that the population of Stockholm sometimes chooses on grounds other than accessibility. In maternity care, pregnant women have had freedom of choice since 1988. The effects of freedom of choice in this sphere were not long in coming. One hospital had to close a ward, while others renovated their facilities and changed their clinical routines (Saltman 1992). Today, freedom of choice in maternity care is well organized, with different hospital profiles and even tours by coach to help expectant mothers (and fathers to be) to choose.

In Stockholm, too, it is noticeable that the hospitals are taking an increased interest in different ways of providing information on their services. Some hospitals have also paid attention to general practitioners, since they refer patients to hospital.

Southern Sweden offers the most choice to individuals when choosing a health care provider, as a result of collaboration between six county councils. In principle, the only exceptions to freedom of choice of provider are certain highly specialized activities in two university hospitals. The participating county councils have also informed their populations more fully about the right to choose. Leaflets have been distributed to households and health care providers, advertisements have been placed in the press, and commercials are run on local television.

The experience in southern Sweden is similar to that in western Sweden and Stockholm. The strongest effects of choice have emerged in the borderland of some county councils, where individuals now have access to the nearest hospital. Accessibility to health care has also been made easier for some commuters. In southern Sweden, it is also possible to recognize winners and losers among the county councils, even though the sums transferred are not as large as in western Sweden. As mentioned above, a change in attitude by providers towards patients constitutes an important parallel effect.

Purchasers and contracts

According to the 1991 Congress of the Federation of the Swedish County Councils, it was still important to maintain a strong political influence in the health services. However, it was clear that elected officials had assumed conflicting roles, as financiers and representatives of providers and consumers, respectively. An important consequence of these discussions about the role of politicians was that many county councils drew up proposals to separate purchasers from providers within the county structure.

Two strategies can be distinguished among the county councils experimenting with new approaches. The first strategy is to focus political activity at the central county council level, and abolish political boards for hospitals and primary care centres. The result is the establishment of a central purchaser. The second strategy is to develop an organization of several smaller sub-county purchasers. Again political boards are in most cases no longer directly responsible for managing hospitals.

As can be seen from the examples of purchaser organizations in Table 1, the solutions vary. At a general level of description, there are points of both similarity and dissimilarity between the county councils. One similarity is that the purchasers comprise a political board of directors and an administrative staff responsible for the concrete negotiations. As a rule, the administrative staff have restricted medical competence. Such competence will be bought

Table 9.1 Examples of purchaser organizations in the county councils in 1993

County council	Purchaser organization
Bohus	14 *local purchasers* are allotted resources according to a population-based model. An administrative staff divided into four offices negotiates with the health care provider
Dalarna	15 *local purchasers* are allotted resources according to a population-based model. Administrative staffs which are to some extent shared negotiate with the health care provider
Stockholm	9 *purchasers at district level* are allotted resources according to a population-based model. Administrative staffs negotiate with the health care provider
Sörmland	The county council board of directors is the *central purchaser*. An administrative staff negotiates with the health care provider

Source: Anell and Svarvar (1994).

externally when necessary and/or be obtained through cooperation with the primary care sector and its general practitioners. Another similarity concerns the tasks of the purchasers, which are based on geographical responsibility for the population. According to this responsibility, as well as problems concerning health status and morbidity, if any, the purchasers shall purchase the care they believe that the population needs. The protection of public health and the coordination of different spheres of medical care – especially primary care and hospital care – are often emphasized as important goals.

There are also important differences between the solutions of the various county councils. This applies not least to the organizational level at which the resources of the purchasers have been pooled. A few county councils – Bohus and Dalarna, for instance – are experimenting with purchasing at a local level. In such cases, the purchasers' responsibility for the population coincides with the boundaries of the municipalities. This approach is expected to improve coordination with the services of the primary municipalities (e.g. social services and care of the elderly) at the same time as strengthening local democratic support. In other county councils, the purchasers have been organized at a district or central level. The most common reason for this is to get away from municipal politics and make it possible to benefit from coordination in the county councils at the same time as the purchasers become more powerful. Administrative costs should also be lower, since the number of contracts will decrease.

Of course, purchasing bodies at both the district and local levels remain accountable to the central political county council management. It is the decisions and rules of the county council that provide the framework for the activities of the purchasers. Furthermore, large capital investments and changes in the overall structure and supply of health services must be approved at the central level.

Even in one and the same county council, there are differences between purchasers' way of working. To a certain extent, this may be related to individual performance, but it can also be the result of different preconditions relating to the accessibility of the care, for example. In Dalarna, there are examples of clear differences between the northern and southern parts of the county council, despite the fact that the fundamental outlines are the same. In the north, which is sparsely populated, the purchasers have placed great responsibility on primary care centres, which have been given a central role as negotiator with the hospitals. In the south, the purchasers have concentrated on exposing activities to competition, and there is less cooperation between purchasers and primary care providers.

The county councils are still learning to work with planned markets. In a handful of county councils, the new principles of control have been introduced into most sectors and have been operating since the early 1990s. The majority of the councils do not have any, or at least very little, direct experience with the new mechanisms. It should be noted that the 1990s have been somewhat unusual, in that the county councils have been in financial difficulties. In the period 1992–94, for instance, the volume of care decreased based on current prices. This in turn meant that the main task of the purchasers was to try to force expenditures down. In most quarters, they have had little influence on the content of care or the coordination of its different spheres. The intention to strengthen democratic accountability and to reduce costs through better coordination with the municipal social services has not yet been realized.

Provider contracts

The restructuring of the county councils in terms of reimbursing providers is to a great extent a matter of doing away with the traditional allocation of resources through global budgets. In their place, contract-like forms of payment between purchasers and providers are being introduced. The development towards extended freedom of choice for the individual complicates these principles of reimbursement, since freedom of choice implies that the money follows the patient.

In the mid-1980s, a few county councils discussed using the DRG system as the basis for prospective payment systems. The major problem was seen to be the negative effects of global budgets on overall productivity. These budgets were often considered unfair by the heads of clinical departments, due to what they felt was a biased incentive structure. Additional activity would result in a budget deficit, which in turn could lead to unpleasant questions from senior hospital management. Fewer procedures, on the contrary, could lead to lower costs than expected, and consequently to rewards. At that time, it was not unusual for hospitals and clinical departments to close down wards and cut back services at the end of the year in order to balance the budget.

Towards the end of the 1980s, productivity in the health care sector came under scrutiny. In this connection, prospective payment systems were looked upon as a method of stimulating increased activity. In 1990, an experiment began in one clinical department in Sweden, the Department of Obstetrics and Gynaecology at Helsingborg Hospital. Subsequently, there has been rapid development in several county councils.

There is a special preference for payments to hospitals in terms of per-case contracts. In general, the principles were first tried out in the surgical sphere. The methods of defining the activities vary (see Table 2). In Dalarna, the local purchasers themselves lay down the principles of reimbursement, in negotiations with the clinical departments (not hospitals). In other county councils which are trying new approaches, the county council's senior management usually establishes the principles that apply to contracts and reimbursement.

Bohus, Stockholm and Sörmland are examples of counties in which practically all hospital care is reimbursed according to prospective case-payment systems. In these county councils, the DRG system forms the basis of the reimbursement for inpatient hospital

Table 9.2 Examples of prospective payments to hospitals for short-term somatic care in 1993

County council	Administrative level which sets principles	Activities in inpatient care	Activities in outpatient care	Cost control
Bohus	County council	DRG (weights developed at Gothenburg hospitals)	Its own classification	Reimbursement of 40 per cent if volume exceeds 101 per cent of the plan; over 104 per cent, no reimbursement
Dalarna	Local purchaser	Mixed depending on service and purchaser	Mixed	Cost-ceilings and/or discounts depending on purchaser
Stockholm	County council	DRG (adjusted Norwegian weights)	Its own classification	Reimbursement of 50 per cent at the most if volume exceeds the plan
Sörmland	County council	Mixed DRG for several surgical spheres (adjusted US Medicare weights)	Its own classification	Reimbursement of 40 per cent if volume of inpatient care exceeds the plan; reimbursement of 10 per cent over 105 per cent less reduction in outpatient care

Source: Anell and Svarvar (1994).

care. These counties have developed separate classification systems for outpatient services. In several quarters, the DRG system is also applied to reimbursement for flows of patients between the county councils. The county council of Gothenburg has produced its own weights for each DRG, which are now used in western Sweden. Other county councils have borrowed the weights from US Medicare or Norwegian applications, and adjusted them to local conditions.

Case-payment systems create incentives to carry out more activities and hence lead to potential problems with controlling total costs. The county councils trying the new solutions are well aware of this. Consequently, in the above-mentioned county councils, the contracts contain additional provisions which stipulate expenditure ceilings. These ceilings have been framed in different ways. In Dalarna, there are negotiated cost ceilings and/or discounts for different services. In other county councils, the reimbursement is gradually reduced if the volume of care exceeds the planned level. In Bohus, for example, reimbursement is reduced to 40 per cent of normal if the actual volume of care exceeds 101 per cent of that planned for. For volumes exceeding 104 per cent, there is no additional reimbursement. Thus, incentives to increase production only apply up to a certain level. Above that level, the incentives change.

As a rule, the decisions on reimbursements are based on average costs within each county. These calculations have in part resulted in large differences between hospitals. There are different strategies for handling the differences, but according to the general intention the reimbursement shall be the same for the same type of care, irrespective of what hospital provides the care. Differences, such as research and teaching in university hospitals, or special undertakings and preconditions, are to be treated separately and paid for through extra funding.

PROBLEMS OF IMPLEMENTATION

It is still too soon to be conclusive about the impact of these new methods of control. However, there is no doubt that accessibility to care has been improved due to increased activity. Several studies also conclude that there is an improvement in productivity. Nowadays, there are few if any queues for hospital treatment; on the contrary, there is overcapacity in some areas. In addition, the political debate on health care has taken a swift turn – from discussions about queues and lack of efficiency to increased understanding of the need to reconsider the structure of the health care system.

It is difficult to analyse the importance of individual measures, since there are many parallel trends. Beyond broad changes at the county council level, many hospitals are also carrying out internal reform programmes. They are reviewing their organizational structure and are introducing new methods to control services and quality. At the national level, several important health care decisions have been made in recent years. Responsibility for care of the elderly has been transferred to the municipalities and a health care guarantee has been stipulated (three months waiting time for ten diagnoses) in order to cope with the queues for hospital treatment (see Spri 1992). Rapid progress in medical technology has also resulted in shorter periods of treatment and more outpatient care. Given these changes, it is hard to isolate the effects of the new models of control at the county council level.

None of the county councils has had much experience with the new systems of control or with patient freedom of choice. In some quarters, there is still a discrepancy between word and deed, which creates further problems. It is not always an easy task to establish what has been discussed, decided and implemented. Where change has been carried through, there are the typical problems of changing from old to new ideas. Existing power structures are being replaced by new ones. Purchasers and health care providers tend to feel uncertain about their new roles. Furthermore, there are often substantial shortcomings in the descriptions of activities and in the cost data. Difficult financial prerequisites (i.e. demands for retrenchments) have also led to demands for overall decisions on structures, parallel to the work on implementing planned markets.

Some specific problems in implementing planned markets from the Swedish point of view are discussed below. The discussion will be limited to overall changes in the county councils regarding freedom of choice for the individual and new control systems in which the separation of purchasers from health care providers is an important guiding principle. First, a lack of information regarding choice will be discussed, together with incentives for elected officials and senior administrators to strengthen the influence of patients. Second, the problem of coordinating freedom of choice for the individual, and the movement towards contracts negotiated by local purchasers, will be considered.

Lack of information regarding choice

Although the county councils have formally adopted the recommendations of the Federation of the Swedish County Councils regarding patient freedom of choice, much of the vital information required to facilitate individual choice is lacking. Several county councils have informed residents and their own staff about the new opportunities for choice; however, these counties have rarely provided factual data that could lead to informed choice.

The results of surveys carried out in a few county councils confirm that there is a need for additional information. Not even in those county councils which have been most active in supplying information do all residents know about their rights to choose. Studies in Stockholm in 1993 showed that every other patient who had been hospitalized did not know that he or she could have chosen another hospital. Apparently, the situation is somewhat better in western Sweden; according to a study undertaken in 1991, 60 per cent of the population knew that they were free to choose a hospital themselves. In southern Sweden, 40 per cent of the population knew that they were free to choose between hospitals and primary care centres (see Anell and Svarvar 1993, 1994).

As noted above, the concrete effects of freedom of choice have varied by geographic region. In general, individuals still 'choose' the nearest hospital and primary health centre as they did before. This is also true when people have to wait for treatment a couple of weeks longer. However, the potential effects of freedom of choice have not yet been adequately examined. For instance, no county council has yet provided residents with data about differences in quality of medical care between hospitals. It is not yet clear whether it would be of interest to the county councils to make individuals well-informed consumers of care.

In those regions where freedom of choice has had an effect on the flow of patients, there is a noticeable change in the behaviour of health care providers as well as of the county councils concerned. Among health care providers, there is increasing awareness that accessibility (e.g. location) is of central importance. These effects are most apparent in western Sweden. For example, a strategically located primary health centre has through county council investments been upgraded to hospital status, with day surgery and increased capacity for diagnostics, so as to better serve the local population which recently – and as a result of the increased freedom of choice – chose to seek treatment in an adjoining county.

Other hospitals are working on improving their equipment and accessibility, which they often emphasize in the information they direct at households and patients. However, at a regional level, practically all concerned agree that there is overcapacity and that some acute-care general hospitals should be closed down, or at least transformed into specialized hospitals. The situation is similar to the 'Prisoner's Dilemma' (cf. Axelrod 1984; Foreman and Roberts 1992). All are agreed that there is a call for joint action, but since they cannot agree on the strategies, they compete instead, which may eventually jeopardize all hospitals. Each single hospital may act in a seemingly rational way, but this will lead to worsening conditions for all parties concerned.

The county councils display an attitude to freedom of choice which is ambiguous. On the one hand, freedom of choice for the individual is a question of importance to the health care politicians. On the other, the effects must not become so great that they become difficult to handle. At least this is true if the outcome is bad for their own county council. Consequently, the incentives to inform residents about their options vary between the county councils depending upon whether they see themselves as potential winners or losers. This also applies to hospitals. In these circumstances, potential losers may try to change the rules of the game. They may threaten to halt cooperation on freedom of choice. In western and southern Sweden, marketing rules have been adopted, which have primarily been forced through by the county councils that have lost out.

In western Sweden, too, there has been disagreement on how to pay for the flows of patients across county council boundaries. As might be expected, the losers hold that the reimbursement should be based on marginal costs, since each single county council should defray the fixed costs. The winners would prefer that the reimbursement is based on full costs. In 1994–95, reimbursement for the flow of patients above a certain norm (based on the previous year) was reduced to 30 per cent of the normal payment.

It is clear that elected officials and senior administrators are now always willing to accept the effects of market mechanisms. As a consequence, they intervene by either changing the rules or using old-style instructions. As has been pointed out elsewhere (Ham and Maynard 1994), the danger of such interventions is too much planning in the market. Transaction costs of the market are added to old bureaucratic structures without any gain in benefits. This problem leads back to the question once raised by Alchian and Demsetz (1972) of who monitors the monitor. Apparently, county councils find it hard to refrain from intervening in the market, if financial and/or political losses become too large. The important question is whether this behaviour is congruent with demands for increased productivity and quality, or if it is nothing more than the most convenient way of handling the effects and avoiding structural change and confrontation. If the latter interpretation is correct, accountability of the monitors (i.e. elected county officials and senior administrators) needs to be strengthened.

Who shall purchase the health care services?

Parallel with the reform of freedom of choice is the development of new models of control based on the separation of purchasers from health care providers. This development runs contrary to freedom of choice for the individual. These two trends also seem to have rather different backgrounds. To some extent, the county councils are trying to introduce simultaneously two models, which are based on different demands regarding planning, follow-up and information. On one hand, many county councils want to establish powerful local purchasers of care for all the residents within its boundaries. On the other hand, they want to give individuals the right to choose hospital and primary health centre.

The difficulty in coordinating these two different approaches is clear. Powerful local purchasers and restricted individual freedom of choice make it essential to study the content of care with regard to the patient's demand for accessibility and good quality services. Otherwise, the patient's perspective could be neglected in negotiations between purchasers and providers of health care about cost-effectiveness (cf. Saltman 1992). In choosing individual freedom of choice as the primary guiding principle, other forms of follow-up and information will be of importance. These include guaranteeing medical quality, which the patient may be less competent to judge, and informing residents on data that can facilitate individual choice. It is also a question of studying the behaviour of health care providers with regard to the limits applied to indications for treatment.

Provided that patients are informed, freedom of choice can lead to competition in accessibility and quality among health care providers. There is no indication, however, that total costs will fall, since the Swedish model of freedom of choice does not provide patients with incentives to choose the most cost-efficient provider. Patient fees are uniform for one and the same level of care. Without adequate follow-up, there is a considerable risk that competition regarding accessibility and quality will gradually be increased, thus eventually resulting in increased total costs.

Where choice is limited in practice (e.g. where there is a lack of alternatives in areas of sparse population), there is no problem with coordinating contracts set up by purchasers and patient freedom of choice, except in principle. Nor is there a problem if individuals for some other reason choose in accordance with the purchasers' contracts. The moment a patient chooses or would like to choose health care providers other than those with whom the purchasers have agreements, the inconsistencies become clear. It stands to reason that this may be the case, especially where local purchasers are established. Given that the money shall follow the patient's choice in the first place, the bargaining position of the purchasers will be so weakened that their entire existence may be questioned.

The county councils have different strategies for the coordination of freedom of choice and purchaser contracts. In some quarters, a reduction in the number of sub-county purchasers is being discussed – partly to make the purchasers more powerful, partly to remedy the inconsistencies between the contracts and the free choice of the individuals. Some county councils are discussing measures that will restrict freedom of choice, primarily by requiring that patients obtain referrals before they can be seen by specialists.

To the external observer, county council reform and the decisions taken by elected officials may appear irrational and to have a double purpose. Health care politicians think it is important to have freedom of choice for the individual, but not to the extent that the patients leave their own county council and take their money with them. At the same time, several county councils have established purchasers who shall purchase care on the basis of the need of a defined area, but not to such an extent that they infringe upon the individual's freedom of choice. From a public choice perspective, however, it may indeed be preferable for elected officials to aim for the best of both worlds, at least as long as the inconsistencies do not become too obvious. But as has been pointed out earlier, such behaviour may in the long run result in higher administrative costs, without any gain in benefits.

No doubt future developments in Sweden will consider how to coordinate freedom of choice for patients and purchaser contracts. Every solution must be considered with a view to the demands of the different interested parties (i.e. consumers, providers and purchasers of health care), and how control over health care services will be divided between them.

SUMMARY

Due to the difficult financial situation and the problems of the traditional system, reform of the Swedish health care sector has been initiated. There are several current trends, some of which are seemingly inconsistent. Above all, this applies to the development towards increased freedom of choice for individuals as well as the establishment of sub-county purchasers who are responsible for those individuals within their catchment area. The inconsistencies have not yet caused insurmountable practical problems. However, information on choice and quality of the alternatives is still insufficient. The county councils which have issued information have in the main reiterated the common message regarding freedom of choice: 'Now you are able to choose'. No county council has produced objective information about the alternatives that would facilitate choice and help create new incentives among providers.

The incentive structure of the county councils appears to give the reform multiple objectives. The county councils have considerable incentives to ensure that structural changes in health services are not too extensive. Health care politicians seem to consider freedom of choice for the individual an important change. The question is to what extent this concern is paying lip-service to the critics of the county councils, and how much it reflects a serious intention to strengthen the power of patients. Some county councils seem to draw back when the problems and their own financial costs become too great. There may be reluctance to inform residents about options, revise reimbursement arrangements so as to change the incentives of the health care providers to receive more patients, or discuss the need for referrals to specialized care.

Free exchange between independent parties is a central market mechanism. In the Swedish health care system, freedom of choice for the individual has represented the predominant market element, and thus has been an important driving force towards increased competition. On the whole, the other changes are based on internal changes in the integrated county council model. It remains to be seen how successful the integrated model will be when it comes to making use of the potential power that freedom of choice and purchaser contracts may represent. All in all, the

Swedish health care sector – twenty-six county councils, each with its own preconditions – will present an interesting platform for further experimentation with different forms of planned markets.

REFERENCES

Alchian, A.A. and Demsetz, H. (1972). 'Production, information costs, and economic organization'. *American Economic Review*, 62, 777–95.

Anell, A. and Svarvar, P. (1993). *Reformed County Council Model – Survey and Analysis of Organizational Reforms in the Swedish Health-Care Sector*. Working Paper 3. Lund: Swedish Institute for Health Economics.

Anell, A. and Svarvar, P. (1994). *Landstingens förnyelse av organisation och styrsystem – är strategierna samordnade*. (Reforms in the Swedish health care sector: Are the different strategies coordinated?). Working Paper 8. Lund: Swedish Institute for Health Economics.

Axelrod, R. (1984). *The Evolution of Cooperation*. New York: Basic Books.

Baumol, W.J., Panzar, J.C. and Willig, R.D. (1982). *Contestable Markets and the Theory of Industry Structure*. New York: Harcourt Brace Jovanovich.

Foreman, S.E. and Roberts, R.D. (1992). 'The power of health care value-adding partnerships: Meeting competition through cooperation'. In S. Levey (ed.), *Hospital Leadership & Accountability*. Ann Arbor, MI: Health Administration Press.

Ham, C. and Maynard, A. (1994). 'Managing the NHS market'. *British Medical Journal*, 308, 845–7.

Lindgren, B. and Roos, P. (1985). *Produktions-, kostnads- och produktivitetsutveckling inom offentligt bedriven hälso- och sjukvård 1960–1980*. (Production-, cost- and productivity development in the public health-care sector 1960–1980.) Stockholm: Allmänna Förlaget.

Petersson, O. *et al*. (1989). *Medborgarens makt*. (Power of the citizen.) Stockholm: Carlsson Bokförlag.

Saltman, R.B. (1992). *Patient Choice and Patient Empowerment: A Conceptual Analysis*. Occasional Paper 40. Stockholm: SNS-Förlag.

Spri (1992). *The Reform of Health Care in Sweden*. Report 339. Stockholm: Spri.

10

COMPETITIVE HOSPITAL MARKETS BASED ON QUALITY: THE CASE OF VIENNA

Christian M. Koeck and Britta Neugaard

INTRODUCTION

Only recently has competition in the health care market centred on competition for quality (Siu *et al.* 1991; Berwick *et al.* 1992). In the past, competing on quality was only one of several competitive behaviours used in the market to improve outcome. Yet quality, as a form of non-price competition, has rarely been utilized as a direct incentive to compete in the health care sector. Consequently, little research has been conducted on the potential uses of competition outside the framework of cost competition (Chirikos 1992). A quality management project underway in Vienna, Austria, is attempting to examine the result of encouraging hospitals to compete based exclusively on quality. Since Austria's health care system operates in a planned market in which, at present, competition based on price is not an issue, quality-based competition can serve as a potential tool to improve the efficiency and quality of care of the hospital system.

The central idea of the project within the Vienna City Hospital Association is to experiment with non-price competition between publicly owned and operated hospitals. It is similar in some ways to the managed competition model currently being developed in the USA, in that it attempts to empower the demand side by providing patients with choice of providers after having been informed of quality outcomes and quality improvement efforts of the existing choices. As with the idea of managed competition, there would be incentives to improve the quality of care by instituting total quality management (TQM) to make quality everyone's business. In theory, the patients in Vienna will be most attracted to those hospitals which provide the highest quality of care.

The general aim of the project is to induce hospitals and nursing homes to compete on the basis of quality and by doing so increase economic pressure on the providers of low-quality care. Studies have shown that competition for quality can influence patient behaviour and demand for hospital services (Chirikos 1992). Competing on quality includes such factors as equipment to please patients or physicians, or to improve the image of the hospital. However, for quality competition to be successful, the measures of quality between the hospitals must be carefully chosen so that a reliable comparison can be made. A population-based approach should be used to compare services delivered, where the experience of the entire target population 'at risk' of using care is represented. Data sources used to compare quality between hospitals include mortality data, patient reports, measures of health-related quality of care, medical records, patient surveys and physician interviews.

In this chapter, we analyse the experience of the Vienna City Hospital Association with quality competition. First, the overall structure of the Vienna hospital system and some aspects of the current health care market are described. This is followed by a description of the history of the project and its underlying strategy. The chapter concludes with a brief discussion of the results to date.

THE PROJECT ENVIRONMENT

The City of Vienna owns and manages twenty-seven acute and long-term care hospitals, including Vienna General Hospital (Allgemeines Krankenhaus der Stadt Wien), the University Medical Centre of Vienna University. These institutions are operated by the Vienna City Hospital Association (VCHA), a public corporation. The basic organizational structure of the VCHA includes a central management and three operational divisions: acute care hospitals, long-term care and nursing homes, and university hospitals. The VCHA provides acute, specialty and long-term care for the Greater Vienna area, serving a population of approximately two million inhabitants. It has a total of 29,000 employees and operates approximately 17,000 beds. The budget for 1994 amounts to approximately US$2.8 billion. The VCHA is by far the largest health care provider in Austria and one of the largest public hospital care providers in Europe.

The VCHA is the only provider of certain inpatient services in Vienna, such as intensive care, mental health and major transplant procedures. In these segments of the market, the VCHA acts as a monopoly. In other areas, the services are provided under intensive competition from other private hospital care providers. These segments of the market include obstetrics and gynaecology, elective surgery, orthopaedics and paediatric care.

Under current financing regulations, hospitals receive their funding from various sources, such as the statutory sickness funds, the federal government, private health insurance and, in most cases the largest portion, from the owner of the hospital. For most Austrian hospitals, these are agencies of state, provincial or municipal governments. In the case of Vienna, almost all these funds are budgeted and distributed according to the amount of care provided, calculated by the number of patient days. This chapter does not cover the conflicting incentives hospitals face under this system or analyse the efficiency questions it raises. But the fact is that hospitals have to attract patients to receive revenue and, in the long run, to stay in business. Therefore, any hospital activity that might attract more patients could ultimately lead to increased economic success. This pressure becomes even more obvious once it is recognized that there is, as in other developed countries, an excess supply of hospital beds. Over the next three to five years, the VCHA will have to close up to 15 per cent of its acute care beds. The first hospitals to be hit by such cuts will be those least attractive to patients and least economically successful.

But hospitals are not only competing for patients in the current hospital environment. They are, perhaps even more aggressively, competing for scarce personnel in certain segments of the labour market. In some hospitals, as much as 30 per cent of all beds had to be closed, due to shortages in skilled labour. The most sought after personnel are nurses and allied health professionals such as laboratory and radiology technicians, physiotherapists and occupational therapists.

Hospitals have strong incentives to compete with each other for patients and skilled labour in order to remain viable contenders in a shrinking and more competitive market. Given the current political situation, public debate and the structural incentives just noted, demands by patients for more customer orientation, personalized services and quality become important. In what constitutes a remarkable difference compared with the situation a number of years ago, these patient demands are increasingly viewed by most hospitals as key factors in a new, more demanding health care environment.

HISTORY OF THE PROJECT

The pilot phase

In 1990, the public in Vienna became increasingly concerned about the quality of care in the city's hospital system. The discussion was triggered by the killing of some thirty elderly patients by four nurses in one of the VCHA hospitals. Following increasing public pressure, the commissioner for health and hospitals had to resign. The new commissioner, together with the leadership of the VCHA, promised sweeping changes in the management of the association and of the individual hospitals. One of the measures considered was the introduction of quality improvement programmes. For this purpose, a department of quality management was founded at the corporate management holding authority of the VCHA. The mission of this new unit was to implement TQM systems in all hospitals and nursing homes of the association.

In a first step, this new unit started a pilot project in a 380-bed general hospital, developing and implementing a TQM model for the hospital. It was based on the experience of successful examples in the USA, described in the literature (Berwick 1989; Laffel and Blumenthal 1989). After 18 months, the project was evaluated and the team was able to present a number of interesting results:

- At this point, the hospitals had little intrinsic motivation or incentive to pursue patient-oriented improvement in quality.
- Customer or patient orientation, one of the major building blocks of TQM, was generally not viewed as an appropriate goal in the context of a public hospital.
- Doctors insisted that quality had to be assessed by experts and not by patients. Therefore, patients' opinions, expectations and satisfaction were not considered to be important in the clinical area and should be limited to the hotel functions of the hospital.
- Most of the quality problems seemed to result from problems in the organizational structure, processes and culture of the

hospital. In general, the evaluation uncovered a considerable lack in inter-professional, interdisciplinary and inter-functional communication and integration. Although this problem has been described repeatedly in the literature, it appeared to be specifically relevant in the context of this pilot site (Georgopoulos 1972; Koeck 1992).

- The commitment of top management appeared to be one of the most important factors for the success of the project.
- Different departments of the hospital differed considerably in their willingness to participate and support the pilot project. Those departments which faced competition in their segment of the health care market seemed to be particularly interested in supporting the quality improvement effort.

Based on these results, the quality management unit of the VCHA concluded that the implementation of a quality improvement project could not be restricted to the mere introduction of a new organizational function. Rather, in line with experience in the USA and in other industries, and following the basic theories of TQM, it would have to be much broader in scope (Garvin 1988). The desired results could only be achieved through a major process of organizational development and change. The goal of this process had to be two-fold. First, to introduce changes into the organizational structure and processes of the hospital, addressing the problems of organizational integration and the lack of cross-functional communication. Second, to facilitate the development of the hospital towards a learning organization, which would increase its ability to continuously change and adapt to the changing environment. For this organizational change process, the hospitals would have to rely on outside consultants and support for a number of years, before they could be able to maintain fully functioning TQM systems themselves. Such support would be provided by the quality management unit of the VCHA, which would take the role of an outside consultant.

The pilot project suggested yet another conclusion. One of the most serious obstacles to the introduction of organizational and managerial change within the VCHA in general and the TQM system in particular, was the lack of competition within the Vienna hospital system and the resulting lack of pressure to change current organizational arrangements. Therefore, it was decided to use the second phase of the TQM implementation project to strengthen competition inside the hospital system around the issue of quality, by linking existing areas of competition with the quality issue.

The decision to use a TQM approach to quality improvement appeared to be the appropriate strategic choice. Based on experience in other areas, the customer focus of TQM could be expected to support the already existing incentive to compete for patients, whereas the employee empowerment aspects of TQM could help institutions become more successful in their competition for skilled labour.

One additional but related area of competition was also introduced. During the second phase of the project, hospitals would have to compete against each other for scarce financial resources for the TQM implementation process, and they would be asked to continuously publish data on their quality improvement activities and the progress of the project.

The contest phase

To achieve these goals, the quality management unit proposed and the senior management of the VCHA should introduce a widely publicized 'Quality Management Demonstration Project at Vienna City Hospital Association'. As a first step, all twenty-seven institutions were asked to participate in a contest to become a 'quality management demonstration hospital or nursing home'. Among the fourteen institutions participating, a board of seven internationally recognized experts in the field of quality management in health care, together with members of the quality management unit, selected six institutions where the implementation of TQM was most likely to succeed over the next three years. The selection was based on the scores of each participating institution on ten criteria. These included an assessment of the commitment of the senior management of the hospital or nursing home, the definition of quality which the institution proposed to use, its view on the role of the patient in the quality improvement process, and the degree to which hospital personnel were involved during the application process.

As an incentive for participation and to facilitate the implementation of the TQM system in a given institution, those selected would, over the three years of the project, receive additional labour resources in terms of trained and qualified personnel, consultancy and support from the quality management unit, extensive training and continuing education for their employees, and the additional funding necessary to build the quality management structure.

Introducing quality competition

To increase the pressure on all hospitals of the system to shift their attention to the issue of quality, those selected as quality management demonstration sites would receive extra funding for public relations activities and would be the focus of the VCHA's advertising and public relation campaigns. The quality management unit, together with professional advertisers, designed an elaborate set of public relations activities, including a quarterly magazine entitled *New Quality*, with reports on the progress of the project and quality improvement within the VCHA in general, as well as reports on a group of prominent supporters of the project from the arts, politics, corporate business and society. The general theme was the notion of the quality management demonstration sites being role models for the 'Vienna Hospital of the Future'. To stimulate competition even further, the VCHA will, commencing in 1996, publish bi-annual data on the quality of care of all twenty-seven hospitals and nursing homes, including data on patient satisfaction, patient orientation, medical outcomes and the process of quality improvement.

Current experience and state of the project

The selection of the quality management demonstration sites was concluded in August 1993. Following the decision, all member institutions of the VCHA were informed of the results of the contest. In a next step, those selected were invited to join negotiations with the quality management unit to draw up a contract, which would be the basis of the working relationship between the quality management unit and the project institutions. These contracts cover the amount of resources received by the hospital for the implementation of the TQM system, the number of staff hours devoted by the institution to quality improvement, quality management training and specific role assignments for the duration of the whole project. The contracts also cover the commitment of the members of the senior managements of the project institutions to participate in at least five days of quality management training and to chair at least one quality circle per year.

All these terms and procedures were considered somewhat revolutionary in the context of a public bureaucracy and in the beginning were met with considerable suspicion. After all the parties involved had gained some experience and had built a basis for trust and commitment, these procedures were increasingly viewed as a role model for improving relationships between member institutions of the VCHA and the central management holding authority.

The actual implementation phase was started in late 1993, after the contracts with the quality management demonstration sites had been signed by the senior managements of the institutions, the presidents of the VCHA and the head of the quality management unit. As a first step, besides commencing the first quality improvement projects, a massive training programme was launched for employees of the project institutions. At the core of this effort is a training curriculum for quality coordinators and quality managers. This programme was designed by the quality management unit and received approval by the Austrian Association for Quality Assurance, a government recognized certification agency for quality systems and training in manufacturing and service industries. This curriculum has, with slight modifications, also been recognized by other major hospital care providers in Austria. The training programmes are open not only to employees of the VCHA, but also to other interested health care workers and other professionals from Austria and abroad. At this point in time, the programme is the only recognized training for quality management in health care and demand for places far exceeds supply. The course has attracted participants from other parts of Austria as well as from Germany.

Although still in its early stages, one can observe signs that the general strategy of linking quality to competition has had some impact. The project hospitals have noted increasing interest by certain personnel groups in vacant positions at the hospitals. Those most sensitive to the new developments are nurses and allied health professions. Moreover, the project itself appears to be attractive for health professionals from different areas. Both the project institutions and the quality management unit have received numerous applications from people wanting to join the project as quality managers and receive special training and continuing education.

Thus far, the public and the media appear to be responding well to the effort. The project has, however, seen several negative developments which need to be mentioned. First (and one sign of the overall success of the underlying strategy), those institutions not included in the project are becoming worried as the public's concern for quality grows and the project institutions receive more and more interest from the media. Most of the hospitals which had either not

applied or not submitted successful proposals are the more powerful and more politically established institutions in the system. For
the senior management of the VCHA and the quality management
unit, this has become an area of increasing concern and attention.
The imminent danger is obvious: with the increasing success of the
project and the overall strategy, the politically powerful institutions
could use their influence to force a political decision leading to the
termination of the project or – and this becomes more likely as the
success of the project becomes more evident – force their inclusion
in the project.

A second problem should also be mentioned. Although at the
time of writing the project is less than a year old, the project staff at
the quality management unit are under increasing work pressures.
Paradoxically, the short-term success of the project might jeopardize the results in the long term. Demand for training and consultancy, and the increasingly complicated logistics of the project, are
requiring increased effort, work and tolerance to stress on the part
of the project staff. To date, the team has only been moderately
successful in dealing with these pressures.

CONCLUSION

To our knowledge, this project represents a new approach to
increase competition inside a publicly operated hospital system. By
shifting the focus from price to quality competition, the project has
tried to utilize the current incentives of the financing system.
Through its focus on a patient-centred definition of quality, it has
attempted to avoid the problems of price competition among
hospitals, which typically lead to an emphasis on improved equipment and those aspects most relevant to physicians. It has also tried
to utilize shortages in certain personnel groups (e.g. nursing and
allied health professions) to introduce, alongside competition for
patients, another area in which hospitals could engage in healthy
competition.

So far, the project seems to be successful in directing the attention
of hospitals to the issue of quality and to inducing the public and the
media to increase their demand for more patient-oriented services.
It has also succeeded in bringing the question of quality to public
attention, focusing the discussion more on the issue of process and
outcome quality and de-emphasizing the notion of structural improvement. The first projects to be completed appear to have had a

positive impact on the personnel involved and of the patient groups which the projects have addressed.

Still, a lot of questions remain to be resolved. Does the project noticeably improve the overall quality of care? Will the benefit from the project justify the cost? Will the project staff be able to maintain the high level of effort over the three years the project is scheduled to run? And will it be able to fight off attempts at the political level to change the underlying strategy or to discontinue the project? These and other questions will have to be addressed in the future.

REFERENCES

Berwick, D.M. (1989). 'Continuous improvement as an ideal in health care'. *New England Journal of Medicine*, 320, 33–6.

Berwick, D.M., Enthoven, A. and Bunker, A.J. (1992). 'Quality management in the NHS: The doctor's role'. *British Medical Journal*, 304, 235–9.

Chirikos, T. (1992). 'Quality competition in local hospital markets: Some econometric evidence from the period 1982–1988'. *Social Science and Medicine*, 34, 1011–21.

Garvin, D.A. (1988). *Managing Quality: The Strategic and Competitive Edge*. New York: Free Press.

Georgopoulos, B.S. (1972). *Organizational Research on Health Institutions*. Ann Arbor, MI: Institute for Social Research.

Koeck, C.M. (1992). 'Wege zur Verbesserung der Qualitaet im Krankenhaus. Zu den organisationstheoretischen Grundlagen der Leistungsqualitaet'. In P. Berner and K. Zapotoczky (eds), *Gesundheit im Brennpunkt*, pp. 45–55. Linz: Veritas Verlag.

Laffel, G. and Blumenthal, D. (1989). 'The case for using industrial quality management science in health care organizations'. *Journal of the American Medical Association*, 262, 2869–73.

Siu, A.L. *et al.* (1991). 'A fair approach to comparing quality of care'. *Health Affairs*, 10, 62–5.

PART IV

CONCLUSION

11

BALANCING SOCIAL AND ECONOMIC RESPONSIBILITY

Richard B. Saltman and Casten von Otter

The chapters in this volume underscore a series of important characteristics about the present, nascent period of development of planned markets in health care systems. They demonstrate equally the degree of commonality across various ongoing activities as well as the dissimilarities that separate different countries' (and sometimes parts of countries') efforts. The chapters also illuminate the issues that still remain to be addressed.

Taken together, the chapters point up the extent to which a broadly historical process of change is underway. Yet the specific character of the model or models which will emerge from the process remains uncertain. Countries like Britain, Sweden and Finland, which traditionally have relied on public structures to deliver health services, have introduced a carefully calibrated set of competitively oriented incentives among their health care providers. While these three countries continue to disagree over the relative importance of alternative market-oriented mechanisms (for example, patient choice as against negotiated contracts), they clearly are in agreement about the advantages of retaining their tax-based, single-payer financing systems. Conversely, countries like the USA and the Netherlands have maintained their efforts to develop a configuration of multiple private insurers (predominantly for-profit and not-for-profit, respectively), convinced that production-side competition alone is not sufficient to achieve the desired policy goals. As noted in the Introduction, the evidence to date suggests that finance-side competition, at least as currently conceptualized, generates more problems than it resolves. A definitive assessment on this question, however, remains some years in the future.

Viewed thematically, the observations that emerge from this book fall into two general groupings: those that concern issues of market structure, and those that reflect issues of health system governance. Each will be discussed in turn.

RESTRUCTURING COMPONENTS OF THE DELIVERY SYSTEM

As reviewed in the Introduction, planned markets in health care systems can take on a variety of conceptual configurations. The chapters in this volume have moved this discussion to a pragmatic level, detailing the diversity of approaches currently underway or under design in different national contexts. Taken together, the chapters suggest a range of specific design concerns that have yet to be adequately addressed or resolved.

A number of key issues revolve around the structure and behaviour of the purchaser. While there is considerable commonality across national lines on the importance of separating purchase from provision, for instance, there is substantial divergence with regard to just who should fulfil this function. Across the different countries, one can find district- or county-level public agencies (the UK and Sweden), sub-county-level public agencies (Sweden), municipal-level public agencies (Finland), private, nationally based companies (the USA and the Netherlands), private, locally based general practitioners (the UK), and private individuals (the USA) all acting as purchasers. One can find examples of planners (Swedish counties), elected politicians (sub-county Swedish boards), nationally appointed trustees (British district health authorities), general practitioners (the UK), commercial businessmen (the Netherlands and the USA) and patients (Sweden) leading these new market arrangements. Further, one finds that they may be purchasing, variously, all hospital services (Finland and Sweden), only elective and diagnostic hospital services up to a specified cost limit (the UK), or both primary and hospital services (the Netherlands and the USA), and that community care (nursing home and home care services) can be added as well (GP fundholders in the UK; the Netherlands).

A related issue concerns who should design the actual contract. For instance, various transaction costs could be reduced through the use of nationally designed model contracts (much as has been done, in a related area, in the development and application of diagnostic-related groups, or DRGs, for case-based hospital payments). Alternatively, a national contract's gains in efficiency and quality standards might be counterbalanced by the impediment it creates to local innovation in contract design. Also, the question of linkage of contract payment to health-related outcomes has only begun to be tackled. In part, the present lack of linkage reflects a generalized difficulty in defining health status and in connecting that status to the outcomes of medical and/or clinical procedures – a connection which is currently the subject of considerable research activity in a number of countries. Yet if such terms as 'contracting for health gain' are to be meaningful, and if public contracting agencies are to be judged on their ability to achieve such gains, this linkage will need to be established.

A related if equally complicated set of structural dilemmas revolves around the role of patient choice in planned markets. There are distinctly different perspectives as to what services patients should be entitled to choose. In some countries, patients have in principle an unrestricted choice of their primary care physician (the UK, Sweden), whereas in other planned markets patients find their choice restricted to those GPs working for a particular municipality (Finland) or private insurer (USA), or with contracts with a particular private insurer (the Netherlands and the USA). In market-oriented models proposed for the Netherlands and the USA, subscriber choice of insurance carrier takes priority over patient choice of provider. Where regional or local public agencies decide service arrangements, citizens can elect those public officials in some countries (Sweden, Finland), whereas they are appointed by the national government in others (the UK). In some planned market models, patients can select a provider outside the geographic boundaries of the public agency district (Sweden) or the private insurer's contracted providers (so-called 'point-of-service' managed care plans in the USA, although with a higher co-payment), while in other models patients are not allowed to receive regular care outside the public agency's (the UK, Finland) or the private insurer's (health maintenance organizations in the USA) stable of contracted health providers. A few municipal governments have begun limited experiments with script vouchers for certain home care and transportation services (Sweden); however, most health services continue to be delivered on a service-tied (the UK, Sweden, Finland, the Netherlands) or benefit-tied (the USA) basis.

Lastly, the relationship between patient choice and aggregate expenditures has been viewed differently within different national health system contexts. In Sweden, for instance, within a framework of fixed public budgets for health care institutions and salaried specialists, patient choice of provider has been viewed as a mechanism which can increase both efficiency and quality without incurring additional costs. In the USA, however, with multiple private insurers and public payers and a predominantly fee-for-service physician payment arrangement, it is the curtailment of patient choice of provider – within limited panels established by managed care plans – which is viewed as the most likely road to increased efficiency and quality, and to expenditure stability.

The chapters in this volume also point up a series of broader questions about the likely future organization and behaviour of planned markets in health systems. One intriguing issue concerns the degree to which market mechanisms generate their own forms of structural change in health care systems, separately if not independently from the political and managerial forces that formally control them. Much as the computer-based information revolution inexorably created the preconditions for re-engineered management practices in private industry, the argument here is that efficiencies of scale in a contract-based provider payment system – generated once again by computer-driven information systems – will push decentralized, firm-based provider systems into one or another form of organizational consolidation. Smee noted that in the UK both GP fundholders and district health authority purchasers are beginning to collaborate and/or merge with their peer organizations in adjacent locales. This process of consolidation raises the possibility that independent GP fundholders, for example, could group together in increasingly larger units to hire professional contract negotiation personnel to handle the non-medical aspects of their activities. Left to develop without state intervention, and combined with the British Government's recently announced decision that all GPs will be expected to become fundholders, this consolidation process could result in a handful of new private corporations that would, together, control most primary care services in the UK. Compared with the pre-reform role of the independent GP, then, the establishment of fundholding practices, instead of increasing the ability of local practitioners to better serve the clinical interests of their patients, could quite oppositely spawn in Britain large corporate primary care companies similar in style and behaviour to for-profit managed care businesses in the USA.

A parallel example of what appear as 'natural' pressures on health system development can be seen in Sweden. Anell recounts the multi-county arrangements for the provision of health services that have now emerged in the large urban areas. The ability of patients in the Gothenburg region to obtain care as they choose within the five participating counties forced painful reductions in facilities and personnel in the outlying counties. The subsequent rationalization of services provoked one of the most affected counties to threaten to end its participation in this regional consortium. Despite the difficulty this county's politicians would have had in justifying such a pull-out to their inhabitants, Anell notes that the other four counties have now agreed to annual volume restrictions on cross-county activity. It will be interesting to see whether these politically established restrictions affect the levels of patients' out-of-county choice. It should also be noted that these five counties have requested approval from the national government to dissolve their current boundaries in order to form a single regional governmental unit. At least two other regions in Sweden have made similar requests. The notion that a larger political body would be better equipped to fund and rationalize existing health care facilities was also one of three proposals for health system reform which the Swedish Minister of Health asked a national committee (HSU 2000) to study in 1992.

Overall, then, while the pressures for larger, more rationally structured provider units initially came in Sweden from a different source (patients) than it did in the UK (contracts), it would appear that a similar set of underlying structural forces is at work within both health care systems. How far these pressures extend, and the degree to which political authorities can in fact respond to them, is a matter of conjecture. Of course, the entire notion of a planned market is predicated on the assumption – as now demonstrated in experience – that market mechanisms can in fact be intentionally harnessed to serve public sector objectives. The presence of unconstrained forces, in a much less tightly regulated health system context, can be observed in the USA, where ownership of provider institutions (particularly acute care and psychiatric hospitals, nursing homes and home care services) has become increasingly concentrated in large corporate structures.

If this notion of inherent health system pressures to consolidate and restructure service provision is correct, it raises complicated questions about the future role of elected public officials in making health care policy. If the underlying nature of health care delivery is changing, such that efficient delivery of care leads to larger but more streamlined provider groups, then the political role becomes one of facilitating inevitable change, and of containing and minimizing whatever negative social consequences might accompany that change. This would suggest that political figures have considerably less ability to steer health policy than they (or we) might like. Of course, political authorities could always halt this process of change through intervention in the planned market process – as Brommels noted was done in western Sweden, and as the emerging oligopoly character of the purchaser market in the UK may soon provoke from Whitehall. Whether such interventions can succeed depends, however, on the real power of the forces they confront – as assessment which cannot be made at this early stage in the reform process.

HEALTH SYSTEM GOVERNANCE

Discussion about the future role of political actors in health systems raises the second, more general question of health system governance. Issues of governance run as an implicit yet powerful thread through the chapters in this volume. Perhaps the most important issues of governance are tied up with one of its key components, the notion of accountability. Accountability remains a crucial dilemma for health systems, reflecting the intrinsic vulnerability of the person who needs care and the absence of the (public and private) third-party payer at the point of service. There are, in principle, four basic forms of accountability. They rely upon (1) the democratic process, (2) market mechanisms, (3) formal regulation and (4) professional ethics (von Otter 1994). Each form receives attention in one or more chapters, underscoring the centrality of accountability as an issue in the design of planned markets.

Democratic processes are perceived as a strong form of accountability. Indeed as Brommels, Anell and on occasion Smee suggest, they may be too strong: a central objective of planned market mechanisms in Sweden, Finland and Britain is to separate elected politicians from the ability to intervene in the day-to-day operation of provider institutions. The balance between what political scientists refer to as appropriate accountability and, conversely, what some economists and medical professionals call inappropriate micro-management, is in practice quite delicate. Of course, removal of elected politicians from direct operating control of providers should not then lead to a complete withering of political accountability in general. As Brommels suggests in Sweden and Finland, the role of local politicians should remain strong as representatives of the citizen and patient. One complicating factor with democratic forms of accountability, however, concerns sanctions. This can be a difficult issue when the breech is of formal rules rather than of political expectations. Indeed, sitting politicians may become more popular with their constituents if they 'break the rules' on their behalf.

Market mechanisms as a form of accountability have to do with choice of supplier, whether by public purchaser, private insurer or individual patient. If a supplier loses the confidence of the buyer, then competitors will take away that suppliers' customers. The sanction involved in this type of accountability, clearly, is financial in character: the threat of smaller market share and lower revenues. As Robinson and Le Grand argue, a sufficient sanction may be only to require the 'contestability' of health care markets and/or 'yard-stick' competition, rather than seeking to establish fully competitive markets in all areas of health service provision. Such a contestability approach, they believe, could lower the transaction cost of market accountability without reducing its potency. The actual effec-tiveness of market-oriented forms of accountability in planned markets is attested to in Anell's description of changes in several Swedish counties and in Harrison's conclusion that planned market mechanisms have affected physicians' practice autonomy in the UK.

Formal regulation is a type of accountability which has been built into the core of most publicly financed and/or operated health systems. As noted in the Introduction, the decentralized character of planned markets requires increased output- rather than input-oriented regulatory activity. As the limiting case on this continuum, private physicians in the USA have long experienced heavy and intensive forms of public regulatory intervention. Arvidsson ex-plores the complex relationship involved in defining the proper regulatory role for government in a planned health care market. The effectiveness of state regulation as a form of accountability remains, however, a somewhat controversial topic. One strong argument for introducing market-oriented mechanisms in health systems has been the inability of command-and-control regulation alone to adequately steer providers towards efficient, effective and patient responsive behaviour. In particular, regulation has tended to be a rather clumsy instrument, reacting belatedly and often to only minor infractions. Moreover, the patterns of provider capture and 'iron triangles' (in which the regulated industry and legislative committees steer the design of the regulatory process) are well-known in both academic and professional circles. Regulatory sanctions, in any event, often require painstaking and expensive documentation and proof.

Ethical accountability based on professional responsibility, self-regulation and trust traditionally has been a major element of health system governance. The key sanction here has been fear of losing professional prestige and esteem in the eyes of one's peers. This form of accountability has been closely associated, however, with a non-commercial or community-oriented understanding of the role of health care providers. In the USA, the growth of physician investment in for-profit health care companies, as well as the continuing commodification of health services in an unfettered market environment, have led to serious concerns about the viability of professional ethics as an effective means of professional self-control and accountability. Harrison documents the noticeable although less dramatic impact in the UK of the emergence of self-governing trusts and fundholding GPs on physician's professional autonomy. The shift among physicians from full self-control towards some mix of managerial and professional autonomy could be viewed as questioning the continued traditional strength of professional ethics as the central arbiter of physician decision-making. A similar problem can emerge from the introduction of economic incentives into the primary care physician's referral decisions. Close evaluation of new planned market mechanisms that give hospital budgets to the primary care level in the UK, Finland and some Swedish counties will be required before concerns about under-referral and under-treatment can be resolved.

Beyond issues of accountability, the chapters here suggest other important topics of governance. A major concern is the link between public purchasers and the population they serve, links which are differently constructed in different planned market models. Robinson and Le Grand note the emergence of a business-style model for district authority purchasers in the UK, with members of the governing boards appointed by the central government, while Brommels reviews the directly elected nature of health care politicians in the Nordic Region. These purchasers also represent geographic areas of different size, ranging from several millions in the larger British districts to populations of only 10,000 in the smaller Finnish municipalities. To the extent that method of selection and size of constituency translate into measures of legitimacy, the different planned market models present ample room for further study and assessment.

An additional issue of governance concerns equity. The chapters do not draw a clear picture as to how the emergence of planned markets will affect the ability of each citizen to obtain the same standard of care. The decentralizing influences suggest that there will be increased diversity of standards of access and quality, certainly for different private insurers in the Netherlands and the USA, but also for more entrepreneurial providers in the UK, Sweden and Finland. This diversity could well be reinforced by reduced transparency of insurer and provider decision-making and data. Alternatively, starting from the perspective of critics who contend that prior health delivery configurations countenanced considerable disparities of service quality and also access, particularly for lower income and/or lower employment status groups, it could be argued that the new planned market arrangements will increase overall consistency across entire national systems, since provider decisions are now likely to be subject to audit and evaluation. Which interpretation is correct will, again, have to await more comprehensive review and assessment.

LOOKING FORWARD

This overview of the progress of planned markets in health care has served to highlight the current diversity among approaches to issues of systems structure and governance. The different approaches presently being pursued within different country contexts underscore the broad nature of the planned market process, as well as the early stage in the development of the process.

Despite this institutional and organizational diversity, however, experience to date with planned market models does allow several broad generic conclusions about the likely future development of these models. Moving beyond the lessons noted in the Introduction, these conclusions suggest the probable direction of the planned market process itself, drawing equally upon activities currently underway as well as those intended and/or proposed activities which the chapters indicate either were not pursued or which were unsuccessful when initially implemented. There would appear to be three main aspects of this developmental process.

The first can be termed 'a return to the middle'. If, to adapt

Hegel's dictum, the prior wholly planned, publicly operated health systems are the thesis, and exclusively market-oriented health system models (on both the finance and production side) are the antithesis, then perhaps the current mixed stage of development of planned markets reflects uncertainty as to exactly where an appropriate social and economic balance lies that could be labelled the synthesis. The mix of state regulation and market incentives found in different models suggests a level of experimentation which is necessary in the real world but which does not fit easily into German Idealist metaphysics. What does appear certain is that earlier, more ambitious attempts to move publicly operated health systems closer to the pure market model – the antithesis – have now drawn back to less dramatic if more institutionally solid 'middle' ground. The notion of creating fully competitive markets among health care providers, for example, is in practice being replaced by less vigorously disruptive notions like contestability and yardstick pricing methods. The original intentions of including large numbers of private as well as public suppliers in these new health care markets have found that privately capitalized providers often require high profit guarantees and that existing public institutions incur lower transaction and other operating costs.

Interestingly, while political debate has focused considerable attention on the more radical demand side, emphasizing the importance of public purchasers (Sweden, the UK) or private insurers (the Netherlands and the USA), the greatest potential for the information-driven efficiencies discussed in the Introduction continues to lie on the supply or provider side of the health market equation. A key aspect of the middle ground on the provider side is the development towards trust-based soft contracts, constructed upon common ownership of the contracting providers or, alternatively, as in private industry, a shared fate among networking private providers (von Otter 1991). The cooperation engendered by soft contracts serves to moderate the impact of economic incentives on socially desirable and responsible arrangements, in contrast to the adversarial and/or litigable character of hard or fixed contracts. Conversely, this movement towards a more socially and economically balanced middle suggests that health care vouchers, also a hard contract with a fixed price, will be less attractive to policy-makers.

Overall, the initial drive during the first few years of the 1990s towards more fully market-oriented versions of planned markets appears to have moderated, leaving in place a larger proportion of the prior health system structures than might have been expected. Moreover, this 'middling' outcome appears to be the result as much of considered debate and experience as it has been due to obstacles to institutional reform. In Sweden, for example, Social Democrats received majorities in all twenty-six counties in the 1994 election, based on their stated intention to preserve service levels in the publicly operated health system (*Dagens Nyheter*, 22 September 1994).

A second developmental element is the importance of developing a new professionalism for physicians. In the golden era of the 1950s and 1960s, medical practitioners were lionized as selfless individuals, often working in solo practice, who received high esteem as well as high rewards from society. By the 1970s, physicians were caught up in critiques of the professional classes levelled by, among others, public choice economists, who view all professionals as little more than self-centred income maximizers (Buchanan 1969; Rothstein 1994). In many respects, while physicians remain well paid, they have lost public respect and credibility. Simplistic views of professional behaviour predominate in some policy-making circles, granting little recognition to the subtleties that accompany the maintenance of a professional orientation, the role of fiduciary ethics, or a certain aspiration to pursue scientific truth. As Harrison illustrates, the role of planned market mechanisms in Britain has been to shift a degree of authority away from physicians towards managers, further reducing the value placed on physicians' independent clinical role.

In contrast to this deconstructionist approach to doctors, a new form of professionalism is needed which can regenerate respect for physicians within knowledge-based hospitals and primary care centres, and where their role in an interactive institutional framework will remain crucial. A major challenge to the shaping of future planned market models will be to develop rewards that fit the physician's essential role in the medical enterprise. Here, too, we can speak of the importance of 'returning to the middle', in which a better balance needs to be achieved.

The third developmental component is the importance of patient choice of health care provider. As suggested above, choice of insurer is an inadequate replacement for patient selection of the medical professionals who deliver their medical services. The ability to select one's provider is a form of bottom-up empowerment that, when matched with the top-down empowerment of electing regional or local health care decision-makers, provides the citizen/patient with meaningful leverage over the character and quality of the health services that are produced. While such choice can be enhanced through better information more widely distributed (Saltman 1994), patients who lack sophisticated medical knowledge can still take appropriate decisions – presuming that all providers are properly vetted and accredited. Planned market models which persist in paternalistic arrangements in which experts decide for patients, or which, as in various managed care systems, substitute manager-led for patient-led provider decisions, may find that their legitimacy in the eyes of patients will erode or even, over time, disappear. In democratic societies in which most citizens can choose the conditions within which they live, it is unlikely that they will permanently accept lack of choice and influence in an area as important to them as health care.

The ultimate test of the planned market era of health reform will, however, be relatively straightforward. It will be the extent to which these reforms facilitate institutional   arrangements better able to provide good quality, accessible health care in a timely and appropriate manner to the citizenry. In the final analysis, planned market models will not be judged on the purity or cleverness of their economic mechanisms, but rather on their ability to resolve specific problems that prevent patients from obtaining the care they need. The theoretical potential will have to be transferred into real-world accomplishments. Ends achieved, not means chosen, will be the basis of assessment.

In a very real sense, economic imperatives will thus have to be balanced with social obligations and responsibilities. This objective will require a renewed appreciation of the importance of trust – 'beyond any contracted function' (Parsons and Smelser 1956) – as a central characteristic of a good health care system. And not only trust of physicians by patients, but trust of providers by other providers. Economic incentives cannot overshadow the essential role of long-term relationships and cooperation among providers as the basis for delivering good health care. Following Williamson (1975), the simplest and most effective way to obtain the necessary levels of trust and cooperation among health providers is through common ownership within the public sector, or if private providers are involved, a clear sense of shared fate. This suggests that planned market models based on soft contracts and patient choice will, over time, prove to be the most resilient of the current arrangements. Not entirely by chance, this view of the future of planned markets corresponds closely to the concept of public competition put forward in an earlier book in this Open University

Press series (Saltman and von Otter 1992). Of course, the appropriateness of any model is conditioned by the national and local context within which it is to be deployed, and no model can fit ideally in every situation and at all times. It will be interesting to note over the next years whether the three developmental parameters just described will in fact become the essential arbiters of a successful planned market model.

REFERENCES

Buchanan, J.M. (1969). *Cost and Choice: An Inquiry in Economic Theory*. Chicago, IL: Markham.

Dagens Nyheter (1994). 'Vården inte till salu med s vid makten', 22 September.

Parsons, T. and Smelser, N.J. (1956). *Economy and Society: A Study in the Integration of Economic and Social Theory*. New York: Free Press. London: International Library of Sociology and Social Reconstruction.

Rothstein, B. (1994). *Vad bör staten göra? Om välfärdsstatens moraliska och politiska logik*. Stockholm: SNS Förlag.

Saltman, R.B. (1994). 'Patient choice and patient empowerment in Northern European health systems: A conceptual framework'. *International Journal of Health Services*, 14, 201–29.

Saltman, R.B. and von Otter, C. (1992). *Planned Markets and Public Competition: Strategic Reform in Northern European Health Systems*. Buckingham: Open University Press.

von Otter, C. (1991). 'The application of market principles to health care'. In D.J. Hunter (ed.), *Paradoxes of Competition for Health*. Leeds: Nuffield Institute.

von Otter, C. (1994). 'Reform strategies in the Swedish public sector'. In F. Naschold and M. Pröhl (eds), *Produktivität Öffentlicher Dienstleistungen*. Gütersloh: Verlag Bertelsmann Stiftung.

Williamson, O. (1975). *Markets and Hierarchies: Analysis and Antitrust Implications*. New York: Free Press.

INDEX

access, 5, 45, 47, 118, 247
  market behaviour and, 9–10
  vouchers and, 136
  see also equalization
accountability, 11, 28, 71–2, 191,
  244–6
  political, 3, 6, 86–7, 88, 91, 95,
  103–8, 216, 222–3, 230, 244–6
  vouchers and, 136, 143–5, 149
actor models, 106, 108
administered price allocation, 77–8
administrative costs, 73
  see also transaction costs
administrative instruments, 10
agency theory, 10, 11, 71
  see also principals
AIDS education, 142
asset management, 41
asset specificity, 35, 36
assets, return on, 180–1
Austria, 227–36
autonomy
  clinical, *see* clinical autonomy
  managerial, 142
  of NHS Trusts, 28

Bartlett, W., 36
Belgium, 46–7, 48
block contracts, 30–3, 36, 39,
  168
BMA (British Medical Association),
  156, 161, 167

boundary crossing, 99, 212–13, 219,
  221, 222, 243
bounded rationality, 34–5
Britain, *see* UK
budget allocation with resource
  pricing, 76–7
business re-engineering, 2–3, 10,
  242

capital, access to, 183
capitalism, 164, 171
capitation, 127–8
capping, 149–50
care, *see* health care
case-based payments, 31, 39, 99, 100,
  107, 241
  see also contracts
catchment areas, 98, 212, 213
centralization, 87
change, 244
  continuous, 189, 204
  market mechanisms and, 242
choice, 3, 29, 142–3, 210, 211–14,
  220–5, 241, 249–50
  and complaints, 144
  contracts and, 29–30
  and expenditure, 242
  fundholding GPs and, 200, 203
  inconsistencies, 224
  information and, 220–1, 225, 250
  limitation, 117, 124–5, 126–7, 130,
  242

choice – *cont'd*
  in managed care, 117, 124–5,
    126–7, 130
  patient awareness of, 21
  vouchers and, 138, 142–3, 146,
    148–50
clinical autonomy, 7, 12, 156–71, 246,
  249
  beneficiaries of, 163–5, 170–1
  control of, 169–71
  and social judgement, 170, 171
coalition politics, 62–3
competition, 1, 4–6, 37–8, 166, 211,
  239–40
  and access, 9–10
  choice and, 223–4
  contestability and, 40–1
  destructive, 222
  fundholding GPs and, 198, 199
  HMOs and, 116
  managed, 115, 227–8
  managed care and, 115–17
  NHS Trusts and, 190
  over selection of patients, 130
  public, 74, 250–1
  quality-based, 227–8, 230–6
  reform and, 90–1
  vouchers and, 137, 149
  yardstick, 38–40, 42, 245
competitive tendering, 5, 15, 28
  see also contracts
complaints, 144
concurrent review, 119, 125–6
consensus, 159
contestability/contestable markets,
  40–1, 198, 212, 245
contract models, 10–12
contracts, 25, 28–34, 39–41, 59–60,
  240–1
  block, 30–3, 36, 39, 168
  cost-and-volume, 31, 32–3, 39, 168
  cost-per-case, 31, 39
  length, 37–8, 42
  politicians and, 101, 103–8
  'preferred providers', 203
  selective, 117, 120, 124, 127
  service, 28
  'soft', 248, 250

  and standards, 169, 223–5
  and transaction costs, 34–6
  types, 30–4, 118, 123, 217–19,
    223–5
  utilization review and, 119
  see also extra-contractual referrals
control, 10, 12, 29, 30, 87, 90, 107,
  167
  of resources, 74–9, 168
cooperation, 250
copayment, 152–3
cost-and-volume contracts, 31, 32–3,
  39, 168
cost-per-case contracts, 31, 39
costs, 12, 15
  competition and, 223–4
  containment, 56–7, 58, 59–60
  information on, 33–4
  managed care and, 129–30
  set-up, 188, 201
  see also transaction costs

Dawson, D., 39–40
de Man, H., 62
decentralization, 8, 87, 90, 105, 247
decision-making, 34–5, 90, 170, 171
Dekker Report, 55–7, 61
demand/supply, 29, 59, 62, 78
democracy, 244–5, 250
  see also accountability
Denmark, 87
dental care, 49, 58, 136, 145
DHAs, *see* district health authorities
diagnostic-related groups, 99, 150,
  217–19, 241
diplomacy, management and, 160,
  162
discrimination, 170, 171
  see also selection
district health authorities, 26, 29–30,
  165–7, 168, 192
  and Trusts, 177–8, 179–80, 184
doctors, *see* physicians
domain theory, 106–7
DRGs, *see* diagnostic-related
  groups

Economist, The, 5, 152

efficiency
    measuring, 40, 161
    transaction costs and, 34–6
    vouchers and, 137, 138, 146–7
    see also productivity
employment conditions, *see under*
        workers
England, *see* UK
entry barriers, 41
equalization/equity, 16, 49, 134, 143,
        198, 247
    see also access
ethics, 163–5, 170–1, 246, 249
evaluation, 6, 7–8, 9, 169, 247
    see also efficiency; monitoring;
        quality *and under* health care
        systems (reforms)
external financing limits, 181
extra-contractual referrals, 30, 31,
        168, 170

fee-for-service systems, 142
FHSAs (family health service
        authorities), 191, 201
finance, 73, 74
    catchment area formulas, 98
    clinical judgement and, 162
    copayment, 152–3
    performance-based, 211
    providers', 141, 142
    vouchers and, 137, 138, 146–7
Finland, 3, 8, 87, 239, 240, 241
    local government and health care
        system, 88, 89–90, 92–8, 101–8
    political accountability, 88, 95,
        97–8
    reforms, 92–8, 101–8
flexibility, 13–14
frame budgets, 90
France, 46–7, 48
franchising, 38
freedom
    clinical, *see* clinical autonomy
    of speech, 156–7, 190
funders, 73, 74
fundholding GP practices, 26–7,
        165–7, 178, 191–205, 242
    and access/selection, 9, 145

extent and growth, 27, 29, 193–5,
        201–3
    finances, 170, 199, 201
    influence, 198, 199–200, 204
    'multi-funds', 202–3
    and preferential treatment, 9

gatekeeping, 99, 118–19, 120, 123–4,
        125, 126, 130, 192
General Medical Service, 191
general practitioners, *see* GPs;
        primary care physicians
Germany, 46–7, 48, 49, 140
goals, 70, 75, 87, 138, 215, 241
Gothenburg, 212–13, 219, 243
governance, 244–7
    see also accountability
government, 71–2, 87, 245
    clinical autonomy and, 164–5
    and costs, 5, 152
    crisis of, 171
    and local government, 91, 105
    and managed care, 116, 130
GPs
    contracts and remuneration, 159,
        191, 193, 197, 198, 204
    freedom of referral, 168
    market-related skills, 204
    waiting time for appointments, 200
    see also clinical autonomy;
        fundholding GP practices;
        primary care physicians
Greece, 46
Griffiths Report, 161–2, 163

health care, and health care systems,
        4–5, 45–50
    clinical criteria for, 60
    commodification, 246
    context, 2, 3–4, 65, 114, 142, 251
    convergence, 68–9, 86, 114, 247–9
    intensity, 141, 151
    managed, *see* managed care
    necessary/luxury, 8–9, 58, 61, 171
    public benefits, 142
    reforms, 1–10
        conflicting objectives, 8–9
        evaluation, 4, 167, 178–9, 247–51

health care, and health care
  systems – *cont'd*
  irreversibility, 16–17
  obstacles to, 7–10
  trends, 68–9, 86, 114
  restricted entry, 48, 53
  tax-based, 46, 48, 239
  two-tier, 135, 202
health insurance, *see* insurance
health insuring organizations
  (HIOs), 121
health maintenance organizations,
  115–16, 117, 118, 119–29
  extent of enrolment, 120, 126
  fundholding GPs and, 203
  group model, 119, 120, 124–5,
    126–7, 129, 131
  and hospitalization rates, 120,
    122–8
  network model, 120, 131
  performance studies, 119–20
  selection by, 145
  staff model, 119, 120, 124–5,
    126–7, 129, 131
Helsinki, 96–7
hierarchical models, 69–70
hierarchy/hierarchical systems, 30,
  69–71, 73
  and efficiency, 33–6, 42
Hiltunen model, 92–3
HIOs (health insuring organizations), 121
HMOs, *see* health maintenance
  organizations
home care, 60, 135, 142, 147
hospitals
  Austrian, 227–36
  cancellations by, 171, 187
  caseload spread over year, 171
  closures, 167, 190, 229
  comparisons, 228
  competing on quality, 227–8, 230–6
  contracting-out by, 5, 15
  Dutch, 55, 60
  finances, 3, 99, 150, 217–19, 222,
    229, 241, 246
  see also contracts
  Finnish, 89–90, 95

fundholding GPs and, 198, 199
internal markets, 177–9
marketing activities, 212–13
mergers, 189
over-capacity, 99, 100–1, 205, 219,
  222, 229
and prescribing, 192
productivity, 99, 129
referrals to, 246
  fundholding GPs and, 192, 195,
    197
  HMOs and, 120, 122–8
  utilization review and, 125–6
shift of services away from,
  204–5
Swedish, 99–101, 212–14, 217–19,
  222
  internal reform programmes,
    212–13, 220, 222
UK, 166, 167
see also Trusts, NHS
US, 145
HSU 2000, 5, 68, 211, 243

Iceland, 87–8
ideology, 15–16, 62–3, 166
  vouchers and, 138, 142, 146, 150
incentives, 1, 37–8, 42, 127–8, 219
  and clinical autonomy, 246
  fundholding GPs and, 192–3, 198
  managed care and, 117–18, 123
  NHS Trusts and, 180, 184, 190
  regulation and, 66
  vouchers and, 137
indemnity insurance, *see under*
  insurance
independent practice associations,
  see IPAs
individual model of health, 164
inequality, 143, 145, 153
  fundholding GPs and, 9, 202
  see also equalization; socio-
    economic status
information
  and choice, 143, 220–1, 225, 250
  decisions not to collect, 179
  and evaluation, 33–4, 141
  on quality, 228

information systems, 2–3, 30,
204
and physician profiling, 118, 125
insurance, 5, 45–50, 239
cost containment, 56–7
and equalization, 49
history of in the Netherlands,
50–63
indemnity, 45, 46, 48–9, 53–4, 60,
115–16
market segmentation, 52–3
market triangle and, 73
risk and, 5, 56–7, 58–9, 127–8, 129
selection in, 116, 129, 145
vouchers and, 151–2
see also health maintenance
organizations
integration, vertical, 37
interests
compatibility, 163–5, 170–1, 192–3,
198, 203
competition and, 222
trade-offs, 191
internal markets, 37, 91, 107, 177–9,
184
fundholding GPs and, 192, 198,
203–4
NHS Trusts and, 203–4
International Symposium on Quality
Management in Health Care,
234
intervention, 66, 71–2, 222–3, 244
utilization review as, 127
IPAs (independent practice
associations), 117, 120, 125, 131
Ireland, 46
'iron triangles', 246
Israel, 135
Italy, 46

James, J., 189
Jarrald, K., 204
Jaures, J., 62
judgement, 34–5, 90, 170, 171
see also clinical autonomy

labour, *see* workers

legislation and regulation, 66, 71, 83
local government, 89–91, 105
see also under Sweden
local health authorities, 202, 203
see also district health authorities; regional health
authorities
London, 190
Luxembourg, 46–7, 48

managed care, 114–31, 242, 250
choice limitation, 117, 124–5,
126–7, 130
and hospitalization rates, 120,
122–8
organizational forms, 120–1
selection and, 129
management, 2, 29, 242, 249
in NHS, 156, 159–62, 183, 188
quality training, 233–4
reactive/proactive, 160, 162,
168–9
see also MBO; TQM
market models, 69, 71–5
market triangle, 73
markets, and market forces/
mechanisms, 1–2, 5, 198, 242,
243
and accountability, 245
contestable, 40–1, 212
effects, 14–17
NHS Trusts and, 180
planned, *see* planned markets
planning and, 11
quasi-markets, 166–7
regulated, 71, 78
MBO (management by objectives),
90, 94, 95, 96
Medicaid (USA), 117, 121, 123–4,
125, 128, 135, 145
medical model, 164
Medicare (USA), 117, 126, 128, 129,
135, 145
mergers, 189, 202–3, 242, 243
Mill, J.S., 62
monitoring, 6, 7–8, 9, 30, 33
of monitors, 222–3
monopolies, 39, 40, 161, 190

NAHAT (National Association of Health Authorities and Trusts), 32
National Health Service (UK), 25–6, 39–40, 177
accountability, 87
clinical autonomy in, 156–71
destabilization, 171
history, 158–63, 169–71
management in, 159–62, 167–9
RAWP formula, 98
reforms, 26–30, 171
evaluation of, 167, 178–9
see also fundholding GP practices; Trusts
needs, 27, 139
Netherlands, the, 4, 46–7, 48, 130, 189, 239, 240, 241
changes in health care system, 49–63
insurance in, 48–63, 145, 151
monitoring process, 8
prioritization criteria, 67
social structure, 49, 50, 52, 53–4, 62
vouchers in, 135, 144–5
New Right, 166
New Zealand, 4, 67, 135
NHS, *see* National Health Service
Nordic countries, 86–108
see also individual countries
Norway, 67, 87

objectives, stated and unstated, 138, 164, 170–1
opportunism, 5, 10, 11, 35–6, 143–4
vouchers and, 151, 152
Oregon list (USA), 67
organizational culture, 119, 126, 127, 130
outreach clinics, 197, 205

paternalism, 250
patients
attitudes to, 212–13
see also quality
choice, *see* choice
clinical autonomy and, 170–1

competence, 141–2, 151
needs, individual/population, 27
opinions/surveys of, 101, 161, 187–8, 199–201, 221, 230
relatives of, and vouchers, 144–5
'Patients' Charter', 187
payment methods, *see* contracts
PCPs, *see* primary care physicians
peer review organizations, 126
pensions, 15
per-case funding, *see* case-based payments
performance indicators, *see* efficiency
physicians, 127, 249
organizational culture, 119–20, 126, 127, 130
primary care, 3, 117–19
see also GPs
profiling, 118–19, 120, 125
socialization, 164
physiotherapy, 58
planned markets, 68–9, 71–2, 74, 78, 166–7, 240, 243
in Austria, 227
and clinical autonomy, 157, 162–3, 165–71
and quality, 250
regulation of, 79–84
resource allocation/control and, 74–9
in Sweden, 211, 220–6
as synthesis, 247–9
see also convergence *under* health care systems
vouchers and, 153
planning, 69–70, 71, 166–7, 201, 222
politicians
elected/appointed, 101, 104–5
professional/voluntary, 105
and professionals, 107, 108
and purchasing, 101, 103–8
and reform, 106–7
responsiveness, 106, 108
role of, 91, 214, 243–4
stakeholder models and, 106
see also accountability
Portugal, 46

PPOs, *see* preferred provider
    organizations
predictability, 138, 139–40, 146,
    147–8
preferred provider organizations,
    117, 120, 124, 125
prescribing, 161, 162, 192, 195, 197
primary care companies, 242
primary care physicians, 3, 117–19
    see also GPs
principals, 10, 11, 71–9
prioritization, *see* rationing
'Prisoner's Dilemma', 222
private practice/services, 9, 47–8, 160
privatization, 10, 13, 91
productivity, 99, 129, 183, 219–20
professional autonomy, *see* clinical
    autonomy
professional bodies, 72
professionalism, 249
profiling, physician, 118–19, 120, 125
PROs, 126
prospective review, 119, 125–6
providers
    financial situations, 141, 142
    self-selection, 119–20
    see also purchaser-provider split
public competition, 74
public firms, 74, 107
purchaser-provider split, 27–8, 29,
    37, 98–101, 108, 165–6, 210, 223
    and clinical autonomy, 168–71
    evaluation of, 167
    forms of, 240
    and managerial orientation, 168–9
    politicians and, 101, 103–8
    see also contracts

quality, 227–8, 230–6, 250
    assessing, 139, 141, 143–4, 228
    choice and, 223–4
    and equalizing access, 47
    fundholding GPs and, 198,
        199–200
    increased diversity, 247
    managed care and, 128–9
    and profitability, 38–9, 143–4
    vouchers and, 149, 150–1

quantity allocation, 75–6
quasi-markets, 166–7
queues, *see* waiting times

rationality, 34–5, 106, 166, 222
rationing/prioritization, 8–9, 67, 164,
    170–1, 203
    clinical autonomy and, 164, 170–1
    vouchers and, 150
    see also objectives
RAWP formula, 98
re-engineering, business, 2–3, 10, 242
redundancies, 3, 15
reform, 2–3, 9–10
    see also under health care
        systems
regional health authorities, 188, 191,
    193–5, 201
regulated markets, 71, 78
regulation, 6, 7–8, 16, 39–40, 65–7,
    83–4, 245–6
    and fundholding GPs, 203
    and NHS Trusts, 190, 191
    of planned markets, 79–84
    reasons for, 66
    in USA, 114
    vouchers and, 153
resource management, 74–9, 168
retrospective review, 119
RHAs (regional health authorities),
    188, 191, 193–5, 201
rights, 62, 79, 80, 81, 156–7, 190
    see also choice
risk, 134–5, 141, 170, 171
    absorption, 134
    evasion, 134–5
    rating, 5, 136, 137
    reallocation, 150
    redirection, 134
    shifting, 134–5, 138, 140, 146,
        150–3
    see also under insurance
rules, 65–7
rural areas, 212, 216, 224
Russia, 135

sanctions, 245, 246
scale, 116, 242

selection
  by hospitals, 145
  competition over, 130
  fundholding GPs and, 9, 145
  in insurance, 45, 116, 129, 145
  managed care and, 129, 145
  vouchers and, 145, 149–51, 152
selective contracting, 117, 120, 124,
    127
self-regulation, 66, 246
self-selection, provider, 119–20
services
  changes in, 183–4, 197–8
  goods and, 139–40
smoking, 170, 171
social thinking, 62–3
social welfare systems, 46, 89
socio-economic status, 48, 53–4, 58,
    136, 143, 247
  see also inequality
solidarity, 16, 45, 47, 57–8, 62–3, 70
  limited, 52, 54–5, 60–3
Spain, 46
stakeholder models, 106–7, 108
Standard Insurance Scheme
    (Netherlands), 54–5, 59, 61
standards, 7–8, 11, 72
  contracts and, 169, 223–5
  see also quality
Stockholm, 209, 213
supply/demand, 29, 59, 62, 78
surpluses, 39, 118, 191, 193, 203
Sweden, 3, 72, 87, 239, 240, 241, 242,
    243
  choice in, 142–3
  equality/inequality, 47–8, 145
  Health and Medical Care Act
    (1982), 209
  Health Services Act (1983), 82, 89
  HSU 2000, 5, 68, 211, 243
  intervention in, 244
  local government and health care
    system, 88–9, 90–1, 98–108,
    209–11, 214–19
  monitoring process, 8
  as planned market, 78–9
  planning in, 82
  political accountability in, 88, 91

  political and economic situation,
    210, 216, 249
  prioritization criteria, 67
  reform in, 90–1, 93, 94, 98–108,
    209–26
  regulation/monitoring in, 8, 81
  vouchers in, 135–6, 137, 140,
    142–3, 146–7
  working conditions in, 13–14
Swedish Medical Association, 210
systems approach, 81–2

tax-based financing systems, 46, 48,
    239
taxi services, 146, 149
tendering, *see* competitive tendering
TQM, 228, 230–6
trade unions, 7, 12–13
transaction cost analysis, 10, 14–15,
    35–8
transaction costs, 9, 10, 34–8, 42, 73,
    222, 240–1, 245
  contract length and, 37–8, 42
  fundholding GPs and, 198
  vouchers and, 146, 152
transportation, 135, 143, 146, 149
triangle, market, 73
'triangles, iron', 246
trust, 11, 246, 250
Trusts, NHS, 28, 39, 166, 167, 177–8,
    179–91, 194, 203–5
  comparison with directly managed
    units, 182, 184–8, 206–8
  and employment contracts, 168
  extent and growth, 181–2, 189, 204
  financial performance, 185
  set-up costs, 188
  treatment figures and patient
    satisfaction, 185, 186–8

UK, 3, 46, 239, 240, 241, 242
  choice in, 143
  equality/inequality in, 9, 47–8
  gatekeeping in, 130
  monitoring process, 8
  as planned market, 78–9
  private practice in, 9, 48, 160
  vouchers in, 135, 143

UK—*cont'd*
  see also National Health Service
unemployment, costs of, 15
universality, *see* equalization
urgency, 169
USA, 4, 6, 14, 113–31, 152–3, 239,
    240, 241, 242
  health care reforms, 113–14
    proposals for, 68, 86, 115, 116,
    131
  hospital mergers, 189
  insurance in, 151–2
  managed care in, 114–31
  managed competition in, 227–8
  ownership structures, 242, 243
  planning in, 113
  prioritization criteria, 67
  as regulated market, 78
  regulation in, 245
  standards in, 8
  vouchers in, 135, 138–9, 145, 147
utilization review, 119, 120, 125–6,
    127, 130

VCHA (Vienna City Hospital
    Association), 227–36
vertical integration, 37

Vienna, 227–36
vouchers, 134–53, 241, 248
  black markets in, 146–7
  copayment schemes, 152–3
  devaluation, 144
  and inequality, 145
  objectives, 137–8, 146
  selection and, 145, 149–51, 152
  suitability of services for, 138–42
  types, 137

waiting times/lists, 60, 143, 169,
    186–7, 199–200, 210, 219, 220
welfare insurance, 15, 134, 138–40
Williamson, Oliver, 34, 37, 250
withholds, 118
workers, 12–15
  disincentives for responsibility, 145
  employment/working conditions,
    3, 12–15, 156–7, 168, 182–3
  freedom of speech, 156–7, 190
  pressures on, 127, 235
  shortages of, 229
Working for Patients, 165, 177, 178,
    203

yardstick competition, 38–40, 42, 245